T0317574

ENERGY STORAGE IN POWER SYSTEMS

ENERGY STORAGE IN POWER SYSTEMS

Francisco Díaz-González
Catalonia Institute for Energy Research, Spain

Andreas Sumper
*Centre d'Innovació Tecnològica en Convertidors Estàtics i Accionaments,
Universitat Politècnica de Catalunya, Barcelona, Spain*

Oriol Gomis-Bellmunt
*Centre d'Innovació Tecnològica en Convertidors Estàtics i Accionaments,
Universitat Politècnica de Catalunya, Barcelona, Spain*

Library of Congress Cataloging-in-Publication Data

Names: Díaz-González, Francisco. | Sumper, Andreas. | Gomis-Bellmunt, Oriol.
Title: Energy storage in power systems / Francisco Díaz-González, Andreas Sumper, Oriol Gomis-Bellmunt.
Description: Chichester, West Sussex : John Wiley & Sons, Inc., 2016. | Includes index.
Identifiers: LCCN 2015044575 | ISBN 9781118971321 (cloth)
Subjects: LCSH: Energy storage. | Electric power systems–Reliability. | Peak load.
Classification: LCC TK2980 .D53 2016 | DDC 621.31/26–dc23
LC record available at http://lccn.loc.gov/2015044575

A catalogue record for this book is available from the British Library.

Cover image: Martin Barraud/Getty

Set in 11/13pt Times by Aptara Inc., New Delhi, India

To our wives and daughters –
Rocío, Sara, Marta, Sofía,
Sílvia, Clara, and Rita.

Contents

Foreword

In response to the move in Europe towards a more sustainable, reliable, and cost-efficient society, European energy policy has set ambitious goals for the European electricity system, fixing the objective of at least 80% decarbonization by 2050.

Distribution networks represent 95% of the electricity grids in Europe. They are therefore a precondition for the retail markets (and also for wholesale) and for the sustainable development of cities and communities, new jobs, and growth.

The transition to a low-carbon society will boost Europe's economy thanks to increased innovation and investment in clean technologies and low- or zero-carbon energy. A low-carbon economy implies a much greater need for renewable energy sources (RES), which are often geographically distributed (90% of the RES in the European Union (EU) are connected to the distribution networks), and also the integration of electric vehicles, which will represent a big shift in demand. Complementary IT solutions are being introduced to electricity networks at both the transmission and the distribution level, adding communication, sensors, and automation to actively manage the new and variable generation and demand. We call these Smart Grid technologies.

Distribution system operators (DSOs) and transmission system operators (TSOs) are responsible for the security of supply and the quality of service on their respective networks. It is EU policy that is driving the need for a reengineering of our electricity networks. New system challenges, including at the distribution level, lead to new network challenges for the pan-European transmission network. Hence, each DSO and TSO in the EU will have to evolve progressively from a "business as usual approach" to a "proactive approach" in order to avoid becoming a bottleneck in the future European electricity system.

It is perhaps surprising that the technologies required to address the new network challenges are, for the most part, not where the research and development (R&D) efforts are most needed. Overall, such inevitable evolution will also require the adaptation of existing regulatory regimes and business models more than technologies.

If the EU is to complete a real internal energy market, regulated companies must play a market facilitation role. TSOs, DSOs, regulators, power generators, retailers, traders, industrial consumers, and storage and RES project developers are all playing key roles in delivering an efficient electricity market. To reach the right setup, however, will involve a multidisciplinary approach to research activities, whereby network operators, manufacturers, and economists must cooperate closely in addressing the many barriers that have been identified - regulatory barriers being an important hurdle to jump.

National regulation still in operation continues to be based on the former design of electricity systems: predictable, controllable, and centralized energy generation, delivering power one-directionally through transmission and then distribution lines, with network charges calculated according to this split.

Now, more and less predictable sources of energy such as wind and solar are being generated locally and connected directly to distribution and sometimes transmission networks (larger plants). This means less controllable generation of energy, the need for bidirectional power flows and the transformation of ordinary consumers to "prosumers." One of the key objectives of network operators, therefore, is to be able to use innovative approaches that are applicable in multivendor environments.

To expand on this last point, the extension and reinforcement of networks in the volume and at the rate required in the lead-up to 2020, 2030, and 2050 will be a costly endeavor. As a result, the DSO–TSO community has identified a number of grid users – demand-side response, electricity storage modules, large consumers, and even aggregated household consumer generators – as potential offerers of what are called system flexibility services, which could, in conjunction with smart technologies, reduce the need for investment in traditional assets.

It is not, however, so easy to make use of such flexibility services under the relatively new laws of the Third Energy (Liberalization) Package, which have imposed a separation of all market activities (generation and retail) and energy networks (transmission and distribution). If one considers this in the context of developing a real market for such services, the need for R&D to address the possible setups and business models becomes ever more apparent. Then there are difficult questions around the funding of R&D and demonstrations of innovative developments under the national regulatory frameworks (not at all available, in many cases).

The next step for the electricity networks R&D roadmap, under development from 2015, is the integration of R&D on storage technology applications into the existing roadmap - storage being able to offer an important form of flexibility. The European Commission is attributing an increasing amount of importance to the integration and alignment of R&D efforts, as well as to its policies in general. This is why, before influencing the calls under the EUs new R&D funding framework to 2020, Horizon 2020, the content of the roadmap is assessed alongside the R&D roadmaps for other energy sectors under the umbrella of the European Commissions EUs Strategic Energy Technology Plan (SET-Plan). The end result is the "Integrated Roadmap," designed as the feed-in document for the Horizon 2020 annual work programs.

Storage is therefore becoming an unavoidable part of the power system, to ensure security of supply and as a crucial form of flexibility.

However, as indicated above, the regulatory frameworks in Europe are not adapted, in the majority of cases, for network operators – and certainly not for DSOs – to integrate storage into their networks; and this despite the considerable economic expenditure being devoted to research by these companies when allowed to do so by the national regulatory authorities.

The costs also remain a main reason for the lack of storage integration in the networks, and are still too high for a strong business case to be made at present. However, in places with a high renewable energies (RES) penetration, storage will be needed whatever the cost. Especially for DSOs, grid-optimized storage can help to address RES peak production and therefore congestion.

The question as to whether network operators will be able to own storage under strict regulation for high-risk and emergency situations, but operated by the market in all normal circumstances, is an issue that is and will continue to be the subject of interesting discussions for some time to come.

Ms. Ana Aguado Cornago
Secretary General of the European Distribution System Operators
for Smart Grids (EDSO)
Brussels, June 2015

Preface

From the outset, the electric power system has been designed to maintain a balance between generation and consumption in real time. This implies severe constraints regarding the short- and long-term operation of the system in terms of security, stability, and the sizing of the units. The current design paradigm is now challenged by the massive rollout of storage units in the power system. In recent years, the electric power system has been undergoing a transition caused by the massive introduction of intermittent renewable generation, which causes a need to incorporate advanced supervision and control features into the classical network operation. With the exponentially increasing numbers of units to be supervised and controlled, advanced computational methods combined with intelligent algorithms will enable the future Smart Grid. Energy storage has not been an initial driver that has triggered the Smart Grid, but it is now definitely a key part of the Smart Grid, not only facilitating the change of technology and design, but also the overlying business models.

The Smart Grid is somehow a starting point that is enabling the massive rollout of storage, leveraging the participation of novel players in the electricity markets who have different business objectives. One important feature of energy storage in power systems is the ability to smoothen intermittent renewable generation, both for large and small-sized operations. The massive rollout of renewables will drive the use of different (centralized or decentralized) storage solutions, which will create a sufficient market size for the storage technology and push the development of the technology.

The origin of this book can be traced back to 2009, when Francisco joined the Catalonia Institute for Energy Research (IREC) to start his doctoral thesis. Andreas and Oriol became his supervisors, and rapidly decided to focus the efforts on the utilization of energy storage technologies in wind power plants. We had gained some experience working in the Centre d'Innovació Tecnològica en Convertidors Estàtics i Accionaments, Technical University of Catalonia (CITCEA–UPC) and IREC on electrical systems and on grid integration of wind farms in some projects with Ecotecnia (which was acquired by Alstom, becoming the wind division of the Alstom group). At that time, we started to move away from the concept of the wind farm to the more appropriate term "wind power plant." Wind power was no longer a fancy green alternative source of energy, which could generate power when the wind blew. It was now part of a massive business, which already bore a very serious level of responsibility in the operation of the whole power system. Transmission system operators were drafting very demanding grid codes, in which wind farms were treated as dependable power plants.

We remember having discussions with some engineers in Ecotecnia (Alstom Renewables, wind division) about the possibility of incorporating energy storage in the wind turbines in order to provide ancillary services. Additionally, these devices could be used for other purposes, as power smoothing, correction of production forecasts, and energy market operations. While the potential of energy storage was evident, there were differing opinions on where to locate the storage, what technology to use, and how to size such energy storage systems. Some engineers supported the idea of wind turbines equipped with energy storage devices that could allow the smooth provision of power adjusted to the forecasted production levels and that could eventually provide ancillary services to the grid. Others argued that it made more sense to operate a single, larger energy storage device at the wind power plant level and provide the same services in an aggregated manner. Other colleagues stressed that eventually energy storage should be deployed on the demand side, close to the consumer, and that it should be combined with demand-side management. Finally, other engineers defended the idea that the optimal solution was to locate the energy storage devices in the distribution substations.

During the realization of the doctoral thesis, some contributions were made on the modeling and control of energy storage systems, especially flywheels combined with wind power plants. Francisco built a scaled test rig with which he could gain some practical experience and demonstrate the possibility of power smoothing using a flywheel. We also realized that there were some impressive advances in the development of energy storage technologies and also on different applications in electric power systems. For example, energy storage was being considered as the only possible solution for preventing rapid power drops in large photovoltaic power plants and in renewable power plants in general. Energy storage was also the backbone of the microgrid concept (which is absolutely necessary to balance power flows) and the lung of the Smart Grid of the future.

By the time the thesis came to an end and was successfully defended in September 2013, we realized that we were starting to understand the potential of energy storage in power systems with a high penetration of renewable energy. Our beliefs regarding the huge potential of energy storage utilization in future power systems triggered the idea of expanding the work done in the doctoral thesis, and in other projects that we had been developing, and start the adventure of writing a book on the topic. At that time, we probably did not appreciate the massive amount of work that was awaiting us when we began the preparation of this book in April 2014.

Let us move forward to spring 2015, at which time we were working to submit the manuscript to the publisher on time. We were writing this preface in the hope and belief that this book could provide some useful guidance to engineers and professionals interested in the utilization of energy storage in power systems that are rich in renewable energy sources. Nowadays, we often hear news stories about paradigm shifts and energy revolutions that will eventually change the way in which we understand electric power in our society. In all these communications, energy storage is part of the equation. We are not certain how future electrical energy systems will be shaped, but we trust that energy storage will play an important role.

According to the scope of the book, its contents are divided into eight main chapters. Chapter 1 first introduces readers to modern power systems. Electric power systems are experiencing a dramatic transformation from the conventional vertically integrated approach with few control actors, towards a system with a high penetration of renewable (and intermittent) generation and, as a consequence, a highly controlled system at any voltage level. As previously noted, such a transformation suggests the introduction of the term "Smart Grid," and this is one of the main concepts underpinning future power system architectures. The Smart Grid architecture is defined in terms of domains, zones, and layers, and these are presented in the chapter. After the presentation of the power system architecture, the chapter continues with the presentation of energy management systems and the fundamentals of power system analysis. In this regard, basic concepts on optimization methods and optimal power flow computational techniques are presented. Viewed together, this results in a didactic approach to an understanding of the fundamentals of power systems. Moreover, though, the chapter also includes a practical example on load-flow calculation.

One of the main drivers of power system transformation is the field of renewable generation, and as such this is presented in Chapter 2. The chapter first discusses the contribution of the various forms of renewable energy in the worldwide energy mix. After this presentation, the chapter classifies the renewable power generation technologies into those based on rotative electrical generators, mechanically coupled to turbines or similar devices (e.g., wind turbines and hydropower); and those based on static power generation sources, producing electricity without any moving devices (e.g., photovoltaics). With regard to the former, the chapter describes wind turbine topologies in detail, and offers two numerical examples on the calculation of the power generated by both fixed- and variable-speed turbines. Finally, with regard to static renewable-based generating technologies, the chapter introduces the concept of photovoltaic generation and proposes a calculation on the analysis of PV panels. The chapter concludes with a brief presentation of the grid code requirements for the grid connection of renewables.

With the stepwise displacement of conventional generating plants by nonsynchronized renewable-based ones, the net level of synchronous power reserves in the system becomes reduced, and this can affect the frequency control in the system. For such reasons, and according to some European grid codes, wind power plants are required to provide power reserves in the same way as conventional generating units.

As a contribution to the description of the requirements for the grid connection of renewables, Chapter 3 presents an extensive literature review on the European grid codes with regard to frequency support. While the chapter looks specifically at wind power plants, the results can be exported to other renewable energy generation technologies. Apart from discussing on grid codes, the chapter includes an extensive literature review on control methods for operating wind turbines, so that they can maintain a predetermined level of power reserves, thus enabling them to participate in tasks related to frequency control.

The three chapters described above serve as a good introduction to electric power systems and renewable generation. These subjects are quite pertinent, and even somehow unavoidable, for a proper understanding of the concepts presented in the rest of the book, which are all centered around energy storage technologies in power systems.

The first chapter on energy storage is Chapter 4. This chapter offers a review of the energy storage technologies that can be potentially included in the electric power system. The chapter covers a great number of technologies, such as pumped hydroelectric storage, compressed air and hydrogen-based systems, secondary batteries, flow batteries, flywheels, superconducting magnetic storage, supercapacitors, and even (although tangentially) the field of thermal storage and the power-to-gas concept. For each technology, the description includes the operating principles, the main components, and the most relevant technical characteristics. The chapter emphasizes the main differences amongst the technologies in a comprehensive manner, including some tables and graphics based on the data collected from several publications and from manufacturers' datasheets. The final part of the chapter discusses power conversion systems for grid connection and the control of storage not synchronized with the network.

Following the description of the technology in Chapter 4, the book tackles the formulation of cost models for the economic assessment of storage technologies. A cost model considering capital, operation and maintenance, replacement, and also end-of-life costs is introduced, based on the literature. The model is demonstrated by means of a numerical example. In this example, the life-cycle costs of different storage systems – both in themselves and while providing various services in the power system – are calculated and evaluated.

The study of the inclusion of storage technologies in the power system usually requires the development of simulation platforms to virtually validate various concepts centered on the design and operation of the technology prior to the commissioning of the system. Accordingly, Chapter 6 presents averaged dynamic models, based on electrical equations, for different storage technologies such as batteries, supercapacitors, and flywheels, as well as for their corresponding power conversion systems. Ultimately, the contents of this chapter can be adopted as a practical approach to the modeling of storage systems. To demonstrate the correctness of the models and of the corresponding control algorithms for the power conversion systems to which the storage containers are attached, the chapter includes various numerical examples. These examples plot the behavior of the storage systems modeled in charge and discharge processes.

In this way, Chapters 4–6 describe the basis for storage technologies and/or offer tools for studies related to the application of the technology. The last two chapters, Chapters 7 and 8, deal specifically with the applications that energy storage systems could potentially provide in the electric power system. Since the power systems of the future will surely be characterized by increasing penetration rates of renewables, most of the storage applications discussed in these chapters are closely related to renewable generation. Chapter 7 presents the potential for short-term applications; that is, for those applications requiring storage in order to rapidly inject or absorb power, over short periods of time, for different purposes. Conversely, Chapter 8 refers to potential mid- and long-term applications: that is, those applications requiring the storage systems to continuously exchange power with the network over periods of hours or even days, for balancing and generation time-shifting purposes.

Both chapters include a numerical example, thus contributing to the practical scope of the book. With regard to short-term applications, a specific example on wind power smoothing with flywheels is offered. This example includes the formulation and solution of an optimization problem, which determines the theoretical optimal operation of the flywheel while providing this service. From the results of this optimization problem, a control algorithm for the flywheel to be executed in real time is derived and also validated using laboratory-scale equipment. Ultimately, the proposed exercise is a good example of the combination of different analytical tools; that is, modeling, optimization, and experimental validation. Finally, the example in Chapter 8 proposes the sizing of a battery bank and its attached power conversion system, building up an isolated power system with PV generation.

For us, writing this book has required tremendous personal effort, but it would not have been possible without the invaluable support received from a number of colleagues, in various forms. We would first like to acknowledge, with thanks, the support received from our colleagues at IREC and CITCEA–UPC: this work is the product of our professional activity over recent years, and throughout this time we have gained experience and knowledge from all of them.

Particularly related to the book, we thank Cristina Corchero and Joana Aina Ortiz for providing us with data for simulations. We thank Jordi Pegueroles and Fernando Bianchi for the design of control algorithms; José Luís Domínguez, Mikel de Prada, and Eduardo Prieto for the figures in Chapter 2; and Gerard del Rosario and Ramón Gumara for the information on laboratory equipment.

In addition, we would like to thank Ms Ana Aguado Cornago for writing the foreword to this book.

Finally, we are also grateful for the permissions received from many authors, institutions, and companies to use figures in the book. In particular, we like to acknowledge the permissions received from IRENA, Redflow Limited, Beacon Power, the World Energy Council, and Knut Erik Nielsen.

We hope that the book will prove to be useful for researchers and engineers. Comments from, and discussions with, readers with diverse backgrounds will be highly appreciated.

Francisco, Andreas, and Oriol
Barcelona, June 2015

1

An Introduction to Modern Power Systems

1.1 Introduction

Power systems are complex structures composed of an enormous number of different installations, economic actors, and – in smaller numbers – system operators. In the traditional approach, the system is dominated by economies of scale. This means that for steadily increasing consumption, a large power generation capacity is installed, mainly nuclear, coal- or gas-fired thermal, and hydroelectric. In order to guarantee the reliability of such a system, a meshed transmission grid at high voltage has to be installed, into which the generators feed. Underlying this transmission system, function of the distribution grid is to conduct the power flow at lower voltage levels to customers, at medium or low voltage. The described power flow is mainly unidirectional, from the generators to the customers, who are connected at medium or low voltage. Only a few customers are connected at high voltage, due to their high loads. Such a system is easy to control, as most of the players (the customers) are passive, only a few actors (generators and system operators) are needed to centrally control the system, and the interfaces are well defined. The most extended economic model in this context is the vertically integrated utility. However, some of the deep fundamentals on which this structure is based can be envisioned, moving from these vertically integrated utilities to the Smart Grid distribution system [1]:

- Economies of scale are no longer applicable to the power system generation, due to the dramatic growth of distributed generation.
- The costs of the various renewable energy technologies have declined steadily due to technological advances.
- Increased environmental concerns on the part of customers and legislators.
- Regulation is enabling the emergence of different players on the electricity market (retailers, energy service providers, etc.)

Energy Storage in Power Systems, First Edition. Francisco Díaz-González, Andreas Sumper and Oriol Gomis-Bellmunt.

These fundamental changes are causing a shift from the vertically integrated approach with few control actors towards a system with a high penetration of renewable (and intermittent) generation and, as a consequence, a system that needs to be highly controlled at all voltage levels. The increasing use of renewable energy not only helps to alleviate fuel poverty, but also promotes decentralized power generation, thereby reducing the dependence on conventional grid-based energy sources. It provides electricity from small-scale generation and microgeneration; working towards reducing the increasing electricity consumption and supplying any surplus generation to the grid. Therefore, microgeneration is a key power generation trend for smart communities, both rural and urban. Distributed generation from micro–combined heat and power (CHP) installations and renewables such as small-scale wind turbines and solar photovoltaics (PV) plays a strong role in this ecosystem. New generation units from renewable energy sources must be established; however, as a result of stochastic generation, those energy resources are intermittent, and possible output fluctuations have to be balanced [2]. Energy storage applications will be used to cope with this problem [3]. All of this leads to the approach to make the grid intelligent: the Smart Grid. A Smart Grid is an electricity network that can intelligently integrate the actions of all of the users connected to it – generators, consumers, and those that do both – in order to efficiently deliver sustainable, economic, and secure electricity supplies [4]. A Smart Grid uses sensing, embedded processing, and digital communications to enable the electricity grid to be observable (able to be measured and visualized), controllable (able to be manipulated and optimized), automated (able to be adapted and to self-heal), and fully integrated (fully interoperable with existing systems, and with the capacity to incorporate a diverse set of energy sources) [5].

One prominent set of actors in modern power systems are "prosumers" ("proactive consumers"). Prosumers are common consumers who become active to help to personally improve or design the goods and services available in the marketplace, transforming both it and their own role as consumers [6]. The strategic integration of prosumers into the electricity system is a challenge. As prosumers are acting outside the boundaries of the traditional electricity companies, ordinary approaches to regulating their behavior have proved to be insufficient. The aggregated potential of flexibility makes the role of the prosumer important for energy systems with high and increasing shares of fluctuating renewable energy sources. To involve different prosumer segments, both utilities and policy need to develop novel strategies. The benefits for prosumers in modern power systems can be summarized as follows:

> **Economic.** The Smart Grid offers the possibility of involving customers, their flexibility being used as an instrument to shed loads and secure stability. It is assumed that customers will allow the distribution system operator (DSO) access to their home automation systems, and that a value chain that links households with the transmission system operator (TSO)

via the DSO will be created in such a way that the flexibility can be used systematically, as can the compensation flowing in the other direction.

Incentives. Incentives may attract customers into a demand–response regime and into distributed energy resources (DER) programs without the need for a proper compensation structure. Poor quality of supply can also be a trigger, especially when there is only one utility operating. Local DER solutions are thus a good option, although the levelized energy costs could be much higher than the supply costs from a centralized utility. Other incentives, such as environmental and social sustainability concerns, comfort, convenience, and so on, could also be drivers.

Technical. Energy storage for electricity is the main key to assuring the stability of a system with intermittent generation, at least for short periods. Ownership models and options for placement in the grid will drive very different solutions. It will be possible for electric cars to supply to the grid (vehicle to grid), which will add to the additional power system storage capability. As long as the distribution operator is in control of, or owns, these facilities, they will be operated in a different manner than if the storage is owned and operated by the community or by a third party working partly on their behalf.

The community. With DER and Smart Grid technologies, communities will gain substantial market power. Traditionally, the utility was in charge of upgrading the infrastructure in order to cater for a sufficient supply capacity and to assure quality. To build a community solution for local supply by means of Smart Grid technologies and DER seems to be the solution for future expansion, at least in rural areas.

Market and trading. New local markets and trading will arise, based on real-time trading, in order to balance the system. The flexibility of customers, local generators, and storage systems will create value on the market to balance the intermittency of renewable generation.

Social. A new form of social cooperation and commitment can be created. For example, customers could start to cooperate to assure that surplus energy that cannot be fed into the system is provided to neighbors and others who are in a position to benefit.

1.2 The Smart Grid Architecture Model

The Smart Grid Architecture Model (SGAM) framework has been developed by the Joint Working Group on standards for Smart Grids, from CEN/CENELEC/ETSI. Its methodology is intended to present the design of Smart Grid use cases by a

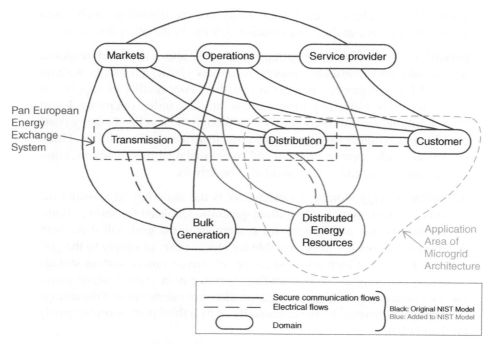

Figure 1.1   The European Conceptual Model, modified from NIST.

holistic architectural definition of an overall Smart Grid infrastructure. Apart from addressing the system architecture through a reference architecture, it also provides an overarching standardization process. The major elements of the described reference architecture are as follows:

1. A high-level framework model (the European Conceptual Model) that is an adapted version of the US NIST (National Institute of Standards and Technology) model, and which bridges between the two models, as shown in Figure 1.1.
2. The SGAM framework as a three-dimensional model with interoperability layers and Smart Grid zones and domains, and that will assist in the architectural design of Smart Grid use cases.
3. Representations of stakeholder views of Smart Grids.

The core of the framework is the Smart Grid plane. In this plane, the power system equipment and energy conversion (electrical processes) viewpoints are linked with the information management viewpoints. These viewpoints can be divided into the physical domains of the electrical and energy processes and the hierarchical zones (or levels) for their management. Figure 1.2 shows the domains and zones of the Smart Grid plane in two dimensions.

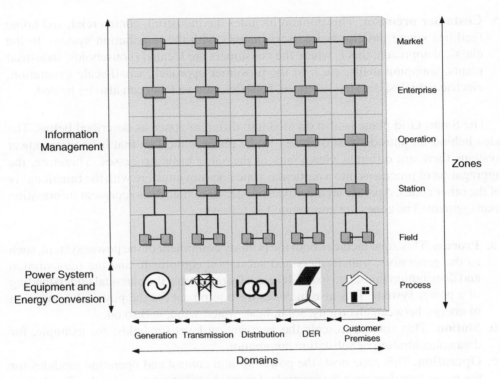

Figure 1.2 The Smart Grid plane.

The different domains represent the power system equipment and energy conversion factors divided into the following subgroups:

1. **Bulk generation**. This domain represents the bulk generation of electricity by power plants. It embodies "classical" power system generation, such as by thermal, nuclear and hydropower plants, as well as large-sized renewable generation such as offshore wind farms and large-scale PV power plants. These facilities are typically connected to the transmission system.
2. **Transmission**. This domain represents the necessary infrastructure and organization for the transmission of large amounts of power over great distances.
3. **Distribution**. This domain represents the necessary infrastructure and organization for the distribution of electricity to the final customers.
4. **DER**. This domain represents generation by means of distributed energy resources, typically using small-scale generation technologies based on renewable energy resources. The range of such generators is typically from 3 kW up to 10 MW; they are connected directly to the distribution grid and can be controlled by the DSO.

5. **Customer premises**. This domain includes the industrial, commercial, and home facilities where the electricity users interact with the distribution system. In the classical approach, this is where the consumers are located (households, industrial plants, shopping malls, etc.). In the prosumer approach, small-scale generation, electric vehicles, demand response, batteries, and so forth can also be hosted.

The Smart Grid plane is also divided into different zones as described below. The idea behind the introduction of zones in the plane is the fact that in modern power systems there are different viewpoints of the same basic processes. Therefore, the aggregation of processes into a particular zone does not interfere with the functionality of the other zones. Apart from the process zone, all of the zones represent information management. The zones are enumerated as follows:

a. **Process**. This zone includes both the primary equipment of the power system, such as the generators, transformers and substation equipment, and the transmission and distribution lines, and loads. It basically represents the classical understanding of a power system. This also includes the equipment for the physical conversion of energy between electricity, solar, heat, water, wind, and so on.
b. **Station**. This zone represents the aggregation level for fields; for example, for data concentration or substation automation.
c. **Operation**. This zone hosts the power system control and operation modules for the respective domains; for example, for the distribution domain, the distribution management system (DMS), for the generation and transmission domain, the Energy Management System (EMS), and so on.
d. **Enterprise**. This zone represents the commercial and organizational processes of an enterprise, including the services (staff training, customer relations management, billing and procurement, etc.) and asset management of the various actors.
e. **Market**. Finally, this zone reflects the possible market involvement of the various actors along the whole production chain.

To complement the Smart Grid plane, five abstract layers are added in order to represent the different viewpoints of the system. This will provide a clear presentation and simple handling of the presented architecture model, enabling the interoperability of the system. Figure 1.3 shows the layer structure above the Smart Grid plane. The different layers are specified as follows:

A. **Components**. In this layer, the physical distribution of all the components participating in the Smart Grid is represented (sensors and actors): system actors, applications, power system equipment, protection and control devices, network infrastructure, and so on.

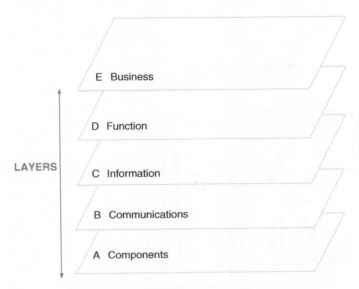

Figure 1.3   The layers of the Smart Grid.

B. **Communication**. This layer describes the communication protocols and mechanisms for exchange of information (connectivity). This layer is important to
   guarantee interoperatibility.
C. **Information**. This layer represents data that are used and exchanged between
   functions, services, and components: data concerning power quality, power flow,
   and protection, from DERs, customers, and meter readings, and so on.
D. **Function**. This layer contains the definition of the functions and services and their
   relationships from an architectural viewpoint. It describes how the functions or
   services are performed independently of actors and physical implementations,
   systems, or components.
E. **Business**. This layer represents the business view of the Smart Grid. It can be
   used to encompass the market, regulatory, and economic structures, as well as
   the policies, business models, products, and services of the market participants.

The SGAM is a very holistic view of the whole Smart Grid architecture, including
widespread viewpoints on the power system. The classical approach to power engineering (i.e., all the domains of the process zone of the component layer) is enriched
by the information management zones and all the other layers. The complementary
zones and layers are defined in order to provide a clear description and to enable interoperability. This does not mean that they did not exist before; they were integrated in
the power system engineering approach. In separating them, the clarity of the solution
is enhanced.

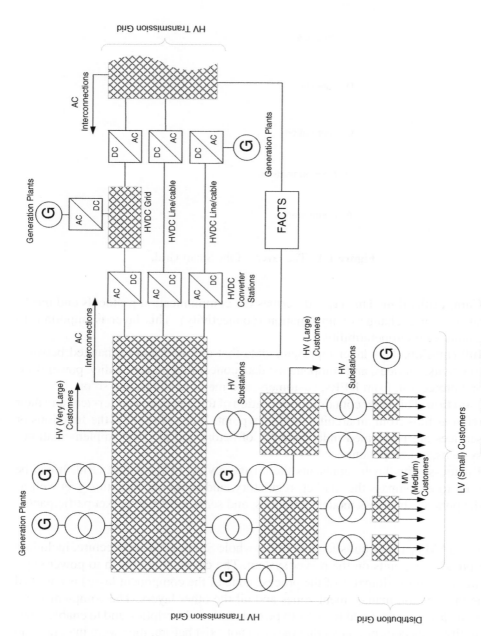

Figure 1.4 The basic structure of a power system.

1.3 The Electric Power System

1.3.1 The Structure of the Power System

The way to organize the electricity system is shown in Figure 1.4. From the SGAM viewpoint, we will focus on all the domains of the process zone of the component layer. The system is divided into generating plants at 6–20 kV (the Bulk Generation domain); a high-voltage transmission grid using 66, 110, 132, 220, 400, 500, and 700 kV (the Transmission domain); the distribution grid, from 3 to 36 kV; the low-voltage grid (the Distribution domain); and consumption (the Customer Premises domain); with the associated protection and control systems. Substations in transmission grids are one of the fundamental components and they have mainly three functions: the first is interconnection of the lines; the second is the power transformation to feed the distribution networks that reach consumers; and, third, they are centers where measurement, protection, interruption, and dispatch operations take place. However, nowadays this structure is complemented by novel approaches to grid structures. Basically, this originated with the need to integrate renewable energy generators at different levels in the power grid. In this sense, there are two approaches that complement the rollout of renewable energy. On the one hand, there is massive connection of small-scale renewable generators at the distribution level (low and medium voltage), mainly photovoltaic installations in households or small-scale generation in light-industrial or commercial facilities (the DER domain). On the other hand, utilities and big investors are increasing their portfolio with renewable generation assets. These plants are at the scale of several megawatts up to hundreds of megawatts. They are considered to operate like traditional power plants and therefore TSOs are asking for compliance with grid codes. In order to intergrade remotely located renewables (in the most common case, offshore wind) in the system, high-voltage direct current (HVDC) technology is employed to connect such plants to the transmission grid (the Bulk Generation domain). HVDC has been employed in the past to perform long-distance transmission using thyristor-based line-commutated converter (LCC) technology. In recent years, voltage-source converter (VSC) technology has been available, which allows better controllability of the link. Nowadays, point-to-point HVDC links using both LCC and VSC technology are the state of the art, and are fixed parts of the power system structure. A logical step is to interconnect the DC lines and create a DC grid in order to increase the reliability of the system.

1.3.2 The Fundamentals of Power System Analysis

The analysis of power systems is fundamentally related to the time horizon with which a certain problem is analyzed. Figure 1.5 shows the relationship between the time horizon and the problem to be analyzed. Table 1.1 shows a small literature sample classified by the time horizon of the problem. We are aware that there are plenty of

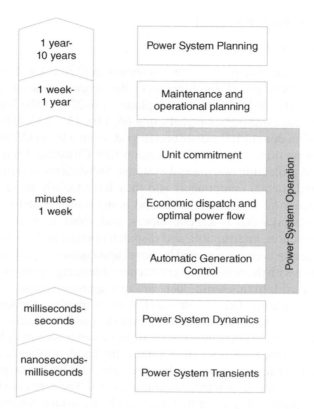

Figure 1.5   A time-horizon perspective of power system studies.

good references that we cannot cite in this book. Readers may take into account the fact that this book is focused on storage application in power systems. In this sense, this chapter aims to provide useful references to literature (see Table 1.1) if readers need to deepen their knowledge of some aspects of power systems.

Power system transients. This type of analysis deals principally with tran-sient events such as switching, faults, and lightning, and covers response frequencies from DC to MHz. The nature of transients is that they have high-frequency components (with rapid attenuation) and also that large voltages and currents can occur. To analyze transients in power systems, it is important to understand their nature through the collection of good data, to form a detailed mathematical model of the system and to solve the resulting coupled differential equations.

Power system dynamics. This type of analysis treats the dynamic proper-ties of electrical machines (synchronous machines), networks, loads, and

Table 1.1 A literature sample on power system modeling and analysis.

Time horizon	Author(s)	Source	Type of modeling or analysis	Reference
Power system transients	L. van der Sluis	*Transients in Power Systems*	Modeling components and analysis of power system transients	[8]
Power system dynamics	P. Kundur	*Power System Stability and Control*	Dynamic modeling and control of power systems	[7]
Power system operation	E. Handschin, A.F. Otero, and J. Cidrás	Steady-state single-phase models of power system components, Chapter 2 in *Electrical Energy Systems*	Modeling of the system components in steady state	[9]
Power system operation	A. Gómez-Expósito and F.L. Alvarado	Load flow, Chapter 3 in *Electrical Energy Systems*	Load flow analysis of power systems	[10]
Power system operation	A.J. Wood, B.F. Wollenberg, and G.B. Sheblé	*Power Generation, Operation and Control*, Chapters 3 and 4	Economic dispatching and unit commitment	[11]
Power system operation	D. Kirschen and G. Strbac	*Fundamentals of Power System Economics*	Electric markets	[12]
Maintenance and operational planning	P. Gill	*Electrical Power Equipment Maintenance and Testing*	Maintenance	[13]
Power system planning	H. Seifi and M.S. Sepasian	*Electric Power System Planning*	System planning	[14]
Power system planning	D. Van Hertem, O. Gomis-Bellmunt, and J. Liang	*HVDC Grids for Transmission of Electrical Energy: Offshore Grids and a Future Supergrid*	HVDC transmission	[15]
Power system operation and dynamics	ABB (formerly Westinghouse)	*Electrical Transmission and Distribution Reference Book*	Transmission and distribution manual	[16]
-	ABB (formerly BBC)	*Switchgear Manual*	Transmission and distribution manual	[17]

interconnected systems on a scale of milliseconds to seconds. It includes control of turbines, power exchange between networks, the behavior of machines in the event of disturbances, transient stability, the equal area criterion, models for small disturbances, voltage control, and the dynamic behavior and control of flexible AC transmission systems (FACTS) devices. The resulting differential equations are solved by analytical or numerical methods, including stability analysis, using approaches from control theory.

Power system operation. This type of analysis includes *steady state analysis*, *automatic generation control* (AGC), *economic dispatch and optimal power flow*, and *unit commitment*. In steady state analysis, basically a stable operation (no time frame is defined) of the power system is analyzed in order to determine voltage drops in the system, loading of the network components, and generator dispatching. AGC is a system for adjusting the power output of multiple generators at different power plants in response to changes in the load on a scale of seconds to minutes. Economic dispatch determines the best way to minimize the overall generator operating costs of a set of generators with differing respective cost functions, over periods ranging from hours to days. The objective of an optimal power flow to combine the power flow with economic dispatch in order to minimize the cost function of the overall system, such as the operating cost, taking into account realistic constraints such as line loads and generation limits. The unit commitment problem involves finding the least-cost dispatch of the available generation resources to meet the electrical load in a long-term perspective.

Maintenance and operational planning. In this type, strategies for the maintenance of electric power equipment are treated, and operational planning for normal and emergency situations is considered. Its time horizon is typically from weeks up to a year. It includes testing (such as of insulating materials and failure modes, and the impact of maintenance on arc-flash hazards) and various maintenance strategies (corrective, preventive, and reliability-centered maintenance).

Power system planning. This is typically carried out over a period of years. As for most planning situations, it is modeled as an optimization problem and the decision-making process is based on both technical and economic considerations. Both generation expansion planning and network expansion planning are treated. With regard to network analysis, steady state analysis techniques (AC or DC load flow, reactive power load flow, etc.) are employed and attention is also paid to security issues.

Security requires special attention in the analysis of power systems. It is defined as the degree of risk in its ability to overcome disturbances (contingencies) without interruption of customer supply [7]. Contingencies refer to outages such as the sudden and unscheduled loss of service of one or more of the main power system components [7]. Usually, power systems are operated according to the deterministic $N-1$ criterion. This means that the permanent loss of one power-system component should not affect the stable operation of the rest of the system. No time frame can be assigned to this kind of analysis.

1.4 Energy Management Systems

An Energy Management System (EMS) is a system that uses computer-aided tools to monitor, control, and optimize the performance of the electric power system. Figure 1.6 shows a schematic of an EMS and its potential application. In utilities, they are used for the generation, transmission, and distribution systems, as well as for monitoring and control functions (where they are usually known as supervisory control and data acquisition, or SCADA, systems). A range of drivers are affecting this development: energy and climate policy, technology in general, and consumer needs. For an EMS, this means local renewable electricity generation such as PV and wind, infrastructure for measuring and managing electricity generation and usage, and new types of energy resources characterized by rapid pivoting and new output profiles. New consumption profiles are a consequence of electric vehicles, induction cookers, and the instant heating of tap water. These developments result in completely new challenges for an EMS in terms of increased dynamics and unpredictability in distribution. Novel needs also include balancing intermittent generation, the difficulty of predict its generation, and capacity problems in distribution systems due to the increased demand. The traditional approach to this type of challenge is to increase the capacity of the generation and power grids. In modern EMS approaches, the optimal operation of such systems can delay these investments.

An alternative approach is to leverage the flexibility of electricity systems by the end users; that is, households, commerce, and industry. The flexibility can come from relocation or reduction of electricity consumption, the use of energy storage, and active management of the generation and conversion of electricity. To achieve this, one must depend on effective decision models for the prediction of electricity consumption (e.g., charging requirements for electric vehicles), electricity generation in buildings and industrial facilities, the monitoring of available flexibility, and optimization models for the utilization of the available sources of flexibility. Furthermore, new technical possibilities are emerging due to the use of energy storage systems (ESSs). Various technologies, such as flywheels, supercapacitors, compressed gas, or battery banks are already being used in the electric power system to offer solutions for peak-load reduction, ancillary services transmission, system reliability, and support

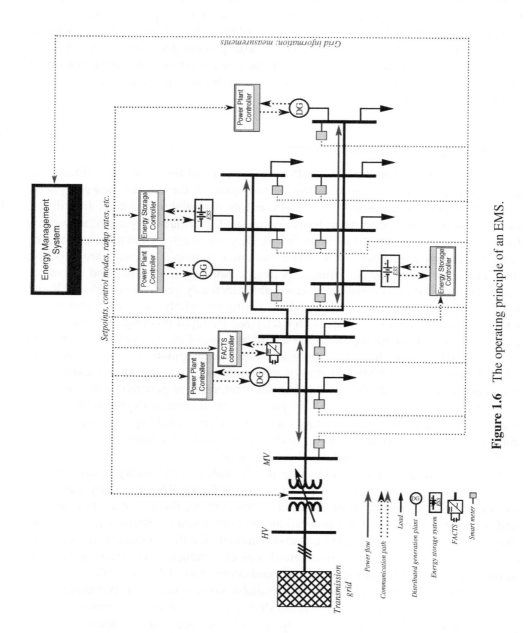

Figure 1.6 The operating principle of an EMS.

for renewables. Moreover, small-ESS technologies are also expected to grow in a distributed fashion along the electric power system, due to the falling price of storage technologies. Such small storage solutions would be mainly based on electric batteries, which would offer additional controllability to the EMS. A large amount and broad distribution of batteries would decouple generation from consumption, opening up new energy management scenarios and opportunities, and also offering new market rules and possibilities for grid operators.

1.5 Computational Techniques

The recent achievements in computational techniques and the scientific developments in rigorous methodologies for the solution of generic problems enable the power system research community to incorporate such technologies to solve specific problems.

1.5.1 Optimization Methods and Optimal Power Flow

Problems in power systems are complex and are usually associated with a large data set. The diversity and versatility of optimal power flow (OPF) formulations does not allow us to find a single optimization technique to solve all OPF problems. Therefore, the solution algorithms that are developed are specifically tailored for the specific problem that needs to be solved. The types of solution algorithm can be divided into deterministic and heuristic methods. Deterministic algorithms provide exact solutions and are designed to guarantee that they will find the optimal solution in an acceptable period of time. However, for very difficult optimization problems (e.g., the so-called "NP-hard" problems), the full resolution increases exponentially with the dimension of the problem. The heuristics do not guarantee an optimal solution; however, they usually find "good" solutions in a "reasonable" amount of time. Generally, heuristic algorithms are very specific and problem-dependent. In order to apply a heuristic algorithm in a problem-independent algorithmic framework, meta-heuristic algorithms have been developed. They provide a set of guidelines or strategies for the development of heuristic optimization algorithms. The deterministic solution techniques for OPF problems are continuous nonlinear programming (NLP), linear programming (LP), quadratic programming (QP), mixed integer linear programming (MILP), and mixed integer nonlinear programming (MINLP). The heuristic solution techniques applied to OPF problems are ant colony optimization (ACO), artificial neural networks (ANN), bacterial foraging algorithms (BFA), chaos optimization algorithms (COA), various evolutionary algorithms (EAs), particle swarm optimization (PSO), simulated annealing (SA), and tabu search (TS). Furthermore, storage elements imply large investments and high operational costs besides the operating constraints, which must be considered in attempts to find optimal solutions. The optimization tools chosen must be tailored to the nature of each case and each system. It is important to emphasize that once storage devices are included in a system, several variables, such as the location, cost, benefit, and charging scheduling, must be also considered in the optimization algorithm alongside the system requirements and constraints.

1.5.2 Security-Constrained Optimal Power Flow

Recent changes in power systems are influencing the way in which they are planned, operated, and controlled. A large penetration of renewable energy generation implies greater uncertainty in systems operation. For instance, in a system with a large penetration of offshore wind, the intermittency of this power generation can compromise its secure operation, as the wind power is not always available when needed to react to an outage, and therefore the accomplishment of the $N - 1$ contingency criterion in some cases (or, better, in some instances) is not guaranteed. The European transmission system is being extended by combining both AC and DC technologies. New devices installed in power grids, such as HVDC or FACTS, increase the controllability of situations but also their complexity. With the increased complexity, the reliability decreases. Therefore, novel methodologies are needed to extend the contingency operations with novel control functionality, taking security aspects into account. Some recent research has been done regarding a method called Security Constrained Optimal Power Flow (SCOPF), which takes the integration of intermittent generation into account. Aragüés et al. [18] analyse the secure and optimal operation of hybrid HVAC–HVDC connected systems with a large penetration of offshore wind, taking into consideration the system's spinning reserves. The operation and economic consequences of requiring higher or lower security are investigated. The SCOPF for hybrid AC–DC power systems allows us to optimize a specified objective function while guaranteeing all the equality and inequality constraints limiting the electrical variables. Security constraints are included, and both the preventive and corrective actions of the TSO can be used to ensure security. Additionally, the SCOPF algorithms must also deal with novel trends in electric power systems, such as the integration of ESSs. Combining renewables and storage could improve grid security and stability. As the penetration of renewables increases, TSOs are asking renewable plants for provide support to the electric power system. As a result, grid codes are starting to regulate generation, imposing technical challenges that could be surmounted through the integration of storage. In such a way, energy storage would allow us to provide voltage regulation, frequency regulation, fault support, ramp rate restrictions, and also power curtailments.

1.6 Microgrids

Microgrids are conceived as self-contained electricity systems with the ability to operate independently of the grid. They could be stand-alone systems; or if tied to the grid, they could be operated by islanding from the grid. Microgrids are also characterized as the "building blocks of Smart Grids." The organization of microgrids is based on the control capabilities over the network operation offered by the increasing penetration of distributed generators. In general, microgrids are an integration platform for supply-side (microgeneration) storage units and demand resources (controllable loads) located in local distribution grids. In the microgrid concept, there is a focus on the local supply of electricity to nearby loads. A microgrid is typically located at the LV level, with a total installed microgeneration capacity below the MW range (with some exceptions). The improvement in storage technologies is also enhancing the horizon regarding microgrid performance. Energy storage is a very important requirement in microgrids, as it allows us to manage energy efficiently but also incorporates new technological possibilities. With the integration of small storage systems in a distributed fashion, close to the point of consumption, new grid structures and topologies can be glimpsed. In this way, these new configurations will lead to new grid concepts, such as prosumers, and also to small grid segments incorporating the storage capacity of electric vehicles into the grid facilities (V2G). Microgrids are normally capable of operating in both grid-connected and emergency (islanded) states. The majority of the future microgrids will be operated mostly in grid-connected mode because of the advantages of bidirectional power interchange. Long-term islanded operation requires large storage sizes and capacity ratings of microgenerators to guarantee the load supply. Demand flexibility and demand response also enable such operation conditions. The difference between microgrids and passive grids penetrated by microsources lies mainly in the management and coordination of the available resources. Operation in islanded mode presents an important challenge, and further research is needed in order to coordinate the power electronic interfaces between resources to guarantee voltage and frequency stability. Usually, hierarchical control at four levels is proposed, based on their bandwidth. The upper level is also called tertiary control or EMS: it is responsible for power flow management and can determine both the active and reactive power, the voltage levels, and the power exchanges with the main grid. The EMS and lower-level controls of the microgrid are responsible for guaranteeing the voltages and power transfer limits inside the microgrid, and therefore the EMS must take the electrical constraints into consideration when it comes to generating power references for the local control units.

1.7   The Regulation of the Electricity System and the Electrical Markets

The electricity markets are described as a very important zone in the Smart Grid plane. Markets are a way of organizing the distribution of commodities in an efficient manner such that the conditions can enhance perfect competition between the actors. However, electricity is not a simple commodity. In order to ensure the reliable and continuous delivery of significant amounts of electricity, the system needs bulk generation plants, transmission and distribution grids, and various control and monitoring functions to maintain the system in a technically feasible state. The simple fact that there is a limited amount of storage capability in the grids (for technical and economic reasons) makes the electricity market unique. The technical differences of the "electricity" commodity have a profound effect on the organization and rules of the electricity markets. Several models of competition have been discussed [12] and are listed below:

Monopoly. This model describes the traditional monopoly utility. In some cases, the utility integrates the generation, transmission, and distribution of electricity; while in other cases, the generation and transmission are integrated in one utility, which provides the electricity to distribution companies that operate within a local monopoly.

Purchasing agency. In this model, independent power producers are integrated into the system, competing between each other and with utility-owned generators. The utility acts as a purchasing agent, buying the best generator offers, and distributes the energy to the customers in a monopoly transmission and distribution system.

Wholesale competition. This model is a hybrid, because there is competition at the generation level but not at the retail level. In this model, distribution companies purchase the electricity directly from generating companies on a wholesale electricity market that takes place mainly at the transmission level. The distribution companies retain a monopoly at the retail level. Large consumers are the exception to this system, because they can purchase electricity directly on the market.

Retail competition. This model allows all consumers to choose their supplier freely. In practice, only large consumers will participate in the wholesale market directly. To enhance the complex participation in the electricity market of small and medium consumers, those consumers purchase their electricity via retailers that are operating in the wholesale market. In this model, the distribution activity is separated from the energy sales to create competition on the retail side, with the objective of reducing electricity prices for consumers. For physical reasons, the transmission and distribution remain as monopolies, regulated by the governmental agencies, and their costs are charged to the consumers.

The introduction of competition in electricity supply has been accompanied by the privatization of utilities. However, privatization is not a condition for the introduction of competition – all of the models described above can also run with public ownership. A market is a mechanism for matching the supply and the demand for a commodity by finding an equilibrium price. Markets can be organized in different ways; each type is complementary to the others and therefore they can be combined. The various types are described as follows:

Spot market. In a spot market, the seller delivers the goods immediately, with no conditions regarding delivery. The buyer also pays for the goods immediately and no party can withdraw once the deal has beendone.

Forward contracts. Forward contracts fix the price and quantity for a future delivery of a commodity, in order to share the price risk.

Future contracts and futures markets. This type is a secondary market, in which the producers and consumers buy or sell forward contracts.

Options. In this type of contract, the contract holder can decide whether or not to make use of the contract. The "call option" gives the holder the right to buy a given amount of a commodity at a price, and the "put option" gives its holder the right to sell a given amount at a specified price.

Given these elementary ways to regulate the electricity system and the available market structures, they can be applied to the electricity system. Due to the particular technical implications of the commodity, the operation of the electricity market is basically organized using the following models:

Bilateral trading. In this type of trading, two partners (a buyer and a seller) agree on a transaction of electricity at a certain price.

Electricity pools. In this type of trading, a centralized pool is created, in which all producers and consumers act. There, all participants are buying and selling electricity, regardless of who might be the final supplier or consumer. This type of trading is well established in power systems, as the transactions and the physical commodity exchange (power flow) are decoupled from each other.

The managed spot market. Electricity systems need to handle imbalances between generation and loads. Therefore, an organized spot market has to be established in order to adjust the daily schedule by means of short-term trading. Since the spot market is the last resort for electrical energy, it strongly affects the other markets.

For secure and reliable operation of the power system, certain "ancillary services" have to provided. These services maintain the quality of the supply in an acceptable range by regulating the frequency, or providing a spinning reserve or power to compensate for imbalances. Typically, these tasks are performed by very flexible generation plants. Also, the TSO could ask for generator schedules to be modified for security reasons, in order to handle the overloading of power lines or transformers. All these commercial transactions have to be settled between all participants and market types as well as with the ancillary services. This process is very complex for the electricity system, and for this reason the settlement system for electricity markets is typically centrally organized.

1.8   Exercise: A Load-Flow Algorithm with Gauss–Seidel

The load-flow study is the most important numerical analysis to determine the flow of electric power at steady state in meshed power systems. This study aims to obtain the magnitude and phase angle of the voltage at each bus, and the real and reactive power flowing in each line for a given load and generation scenario. There are several numerical algorithms that solve such problems [19]; for example, Gauss–Seidel, Newton–Raphson, and the fast decoupled method. In this exercise, a given hypothetical grid (Figure 1.7) should be solved by means of computer programming, applying the simplest load-flow method, namely Gauss–Seidel. A simple case will be treated: a six-bus system with only one slack generator, with the rest of the buses being PQ nodes. The following solution has been written in MATLAB® and is presented in a number of steps:

1. For given bus data, a function is created that shows the data in "per unit" (pu) related to the buses of the system, specifying the bus number, the bus active power generation, the bus reactive power generation, the bus active power load, and the bus reactive power load, respectively:

```
function data = busdata()
%               | Bus |  PGi | QGi | PLi   | QLi  |
data =     [      1     0.0    0       0      0;
                  2     0.8    0.4     0      0;
                  3     0.7    0.3     0      0;
                  4     0.0    0       0.6    0.3;
                  5     0.0    0       0.5    0.2;
                  6     0.0    0       0.9    0.5];
```

2. Next, the power lines or transformers are identified by the following function containing the line data matrix. First, the bus numbers from where the line is coming and to where the line is going are identified. Then, the resistance, reactance, and susceptance of the line, in pu, are introduced:

```
function data = linedata()

%          | From |  To  |   R  |   X  |  B/2 |
%          | Bus  | Bus  |      |      |      |
data    = [ 1      2       0.10   0.20   0.02;
            1      4       0.06   0.20   0.02;
```

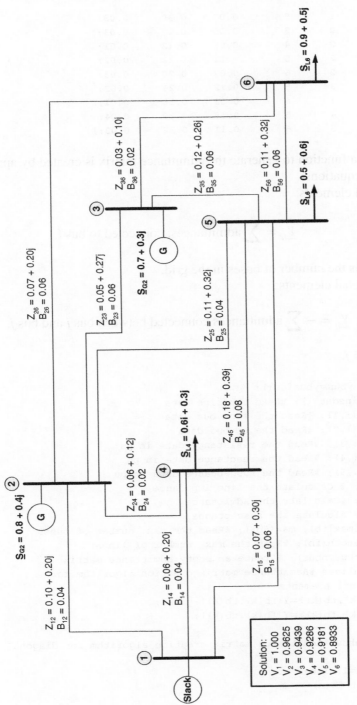

Figure 1.7 The six-bus system for the load-flow study.

1	5	0.07	0.30	0.03;
2	3	0.05	0.27	0.03;
2	4	0.06	0.12	0.01;
2	5	0.11	0.32	0.02;
2	6	0.07	0.20	0.03;
3	5	0.12	0.26	0.03;
3	6	0.03	0.10	0.01;
4	5	0.18	0.39	0.04;
5	6	0.11	0.32	0.03;];

3. After that, a function to generate the admittance matrix is created by applying the following equations:
   - Diagonal elements:

$$\underline{Y}_{ii} = \sum_{i=1}^{n} \text{admittances: connected to bus } i$$

   where n is the number of buses in the grid.
   - Off-diagonal elements:

$$\underline{Y}_{ij} = -\sum \text{admittances connected between bus } i \text{ and bus } j.$$

where $i \neq j$.

```
function ybus=ybusfct();
lines= linedata(); %Read the line data
fb=lines(:,1); %Read the from bus data
tb=lines(:,2); %Read the to bus data
r=lines(:,3); %Read the resistance data in pu
x=lines(:,4); %Read the reactance data in pu
b=lines(:,5); %Read the susceptance data in pu
zl=r+i*x; % Calculate the line impedance
yl=1./zl; %Calculate the admittance
bl=i*b; %Calculate the susceptance
nbus=max(max(tb), max(fb)); %Read the max. number of buses
nlines=length(fb); %Read the max. number of lines
Y=zeros(nbus,nbus); %Create an empty admittance matrix
for k=1:nlines %Admittance matrix creation algorithm for
off-diagonal elements
    Y(fb(k),tb(k))=Y(fb(k),tb(k))-yl(k);
    Y(tb(k),fb(k))=Y(fb(k),tb(k));
end
for m=1:nbus %Admittance matrix creation algorithm for diagonal
elements
    for n=1:nlines
        if fb(n)== m
```

```
            Y(m,m)=Y(m,m)+yl(n)+b(n);
        elseif tb(n)== m
            Y(m,m)=Y(m,m)+yl(n)+b(n);
        end
    end
end
ybus=Y;
```

4. Calculation of the voltages from the given data, using the Gauss–Seidel method. The voltage of each bus for iteration $m + 1$ can be calculated from the active power P_i and the reactive power Q_i, the admittance matrix values, and the voltages from the iteration m by

$$\underline{V}_{i(m+1)} = \frac{P_i - jQ_i}{\underline{Y}_{ii}\underline{V}_{i(m)}^*} - \sum_{k \neq i}^{n} \frac{\underline{Y}_{ik}}{\underline{Y}_{ii}} \underline{V}_{k(m)}, \qquad (1.1)$$

$$i \qquad\qquad = 2 \dots n.$$

The value of the starting voltages for buses 2–6 is assumed to be 1 pu (flat start). For simplification of the algorithm, the voltage calculation can be performed by

$$\underline{YV} = \sum_{k \neq i}^{n} \underline{Y}_{ik}\underline{V}_{k(m)}$$

$$\underline{V}_{i(m+1)} = \frac{\dfrac{P_i - jQ_i}{\underline{V}_{i(m)}^*} - \underline{YV}}{\underline{Y}_{ii}}, \qquad (1.2)$$

$$i = 2 \dots n.$$

The following code is the main program that has to be executed in the same folder where all the other functions are saved:

```
ybus=ybusfct();
bdata = busdata(); %Read the bus data
bus=bdata(:,1); %Read the buses
GenP=bdata(:,2); %Read the generation active power
GenQ=bdata(:,3); %Read the generation reactive power
LoadP=bdata(:,4); %Read the load active power
LoadQ=bdata(:,5); %Read the load reactive power
nbus=max(bus); %Read the number of buses
P=GenP-LoadP; %Calculate the active power injection
Q=GenQ-LoadQ; %Calculate the reactive power injection
```

```
V=ones(nbus,1); %Initialvoltages. Flat start at 1\,pu
Viter=V; %Storing the voltage of the current iteration
tol=1; %initial value of the tolerance
iter=1; %number of iterations
while ((tol> 0.00001)& (iter<100))
    for i=2:nbus;
        YV=0;
        for k=1:nbus %Calculation of the sum of YV
            if i ~ =k
                YV=YV+ybus(i,k)*V(k);
            end
        end
        V(i)=((P(i)-j*Q(i))/conj(V(i))-YV)/(ybus(i,i)); %Gauss Seidel
        algorithm
    end
    iter=iter+1; %Iteration counter
    tol= max (abs(V-Viter)); %Tolerance calculation
    Viter=V; %Storing the result of the current iteration
end
display(iter);
display(abs(V));
```

An additional function is provided to convert the polar coordinates of the complex number into rectangular coordinates:

```
function rect = pol2rect(mag,angle)
rect = mag*cos(angle) + j*r*sin(angle);
```

The resulting voltages are obtained after 38 iterations and are displayed in Figure 1.7 (from bus 1 to bus 6, in pu).

2

Generating Systems Based on Renewable Power

2.1 Renewable Power Systems

Renewable power has been used by humankind for several millennia. In fact, until the Industrial Revolution, renewable power (including hydropower, wind power, biofuels, and solar power) was actually the only energy source available. The Industrial Revolution in the eighteenth century led to the dawning of an era of a new generation of energy sources, which spawned great developments for civilization, but also caused (some decades afterwards) a tremendous increase in the planetary pollution levels and in global warming. One of the side effects of the Industrial Revolution was that renewable energy was abandoned, since it was far less practical and powerful than the "new" machines powered by coal.

Some decades later, the development of electrical engineering in the nineteenth century allowed engineers to design systems that were able to generate, transmit, and distribute electrical energy to the loads. The first electric power plants were powered by coal or water. Later, fossil fuels derived from oil and nuclear fuels were increasingly used.

In recent decades, we have realized that we need to rethink the way in which we humans generate and use energy in order to reduce pollution, halt planetary global warming, and mitigate the risks of nuclear accidents. The scientific evidence of such problems has created alarm in society, which in turn has pushed decision-makers to take strong actions. These actions have facilitated the development of forms of renewable energy, mainly hydroelectric, wind power, and photovoltaic (PV) solar power.

Actual power systems have an increasing share of the renewable energy sources. While some decades ago renewable energy was considered as alternative energy, nowadays there is a massive industry in the field of renewables, and the energy mix of most developed and developing countries shows a substantial penetration of

Energy Storage in Power Systems, First Edition. Francisco Díaz-González, Andreas Sumper and Oriol Gomis-Bellmunt.
© 2016 John Wiley & Sons, Ltd. Published 2016 by John Wiley & Sons, Ltd.

renewable energy. Such increased penetration presents a number of technical and economic challenges related to the integration of such renewable energy sources into the grid. The uncontrollability of these resources has motivated the development of advanced operational strategies for ensuring the safe and secure operation of power systems, thus giving a boost to a profound change in the development of modern power systems, which need to include power electronic converters, flexible and manageable demand, energy storage, intelligent communication networks, and advanced controllers.

According to the International Renewable Energy Agency (IRENA) [20] (Figures 2.1 and 2.2), in 2013 the cumulative worldwide installed capacity of renewable generation was above 1600 GW, where 1138 GW corresponded to hydropower plants, 311 GW to onshore wind, 7.5 GW to offshore wind, 136 GW to PV plants, 3.4 GW to solar thermal, 73 GW to solid biomass, 12 GW to biogas, 11.7 GW to geothermal, and 0.5 GW to tide, wave, and ocean power. In Figure 2.3, it is shown that China is leading worldwide in terms of installed capacity, followed by the United States (US), Brazil, and Germany. It is clear that the development of renewables is being experienced worldwide, not only in some specific countries. According to the World Energy Council [21], future trends (Figure 2.4) show an increasingly important role for solar thermal power plants, also with a strong development of onshore and offshore wind.

An important feature of renewable energy is that the capacity factor (which can be defined as the ratio of the net MWh of electricity generated in a given year to the electricity that could have been generated at continuous full-power operation) tends to be much lower than the capacity factor of conventional generation sources. This is due to the fact that the resource is not always available (clearly for wind and solar energy). Figure 2.5 shows that for wind or solar power, a given power capacity leads to a more reduced level of energy production.

The cost of the energy produced by renewable energy sources (RES) has been decreasing in recent decades. A study from the World Energy Council [21] analyzes the cost of the different energy sources, including a comprehensive comparative study of the costs of producing electricity from a wide range of conventional and nonconventional sources. The study provides reference costs based on real project information, focusing on the leading forms of renewable energy (wind, solar PV and solar thermal, marine, biomass, hydroelectric, and geothermal) and conventional technologies across different geographical locations worldwide. The costs considered include the capital expenditure (CAPEX), which includes the total cost of developing and constructing a plant, excluding any grid-connection charges, the operating expenditure (OPEX), which includes the total annual operating expenditure, and the capacity factor [21]. The various technologies are compared using the levelized cost of electricity (LCOE), which can be defined as the US$/MWh value that represents the total life-cycle cost of producing a MWh of power using a specific technology.

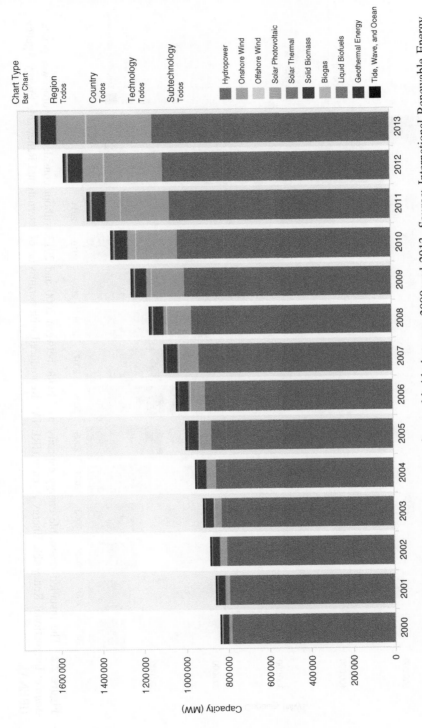

Figure 2.1 The installed renewable energy capacity worldwide between 2000 and 2013. *Source*: International Renewable Energy Agency (IRENA). Reproduced with permission of International Renewable Energy Agency (IRENA).

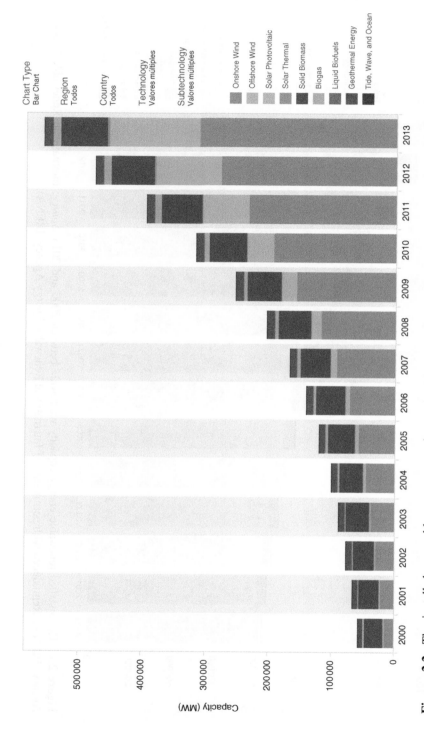

Figure 2.2 The installed renewable energy capacity worldwide between 2000 and 2013 without considering hydropower plants. *Source*: International Renewable Energy Agency (IRENA). Reproduced with permission of International Renewable Energy Agency (IRENA).

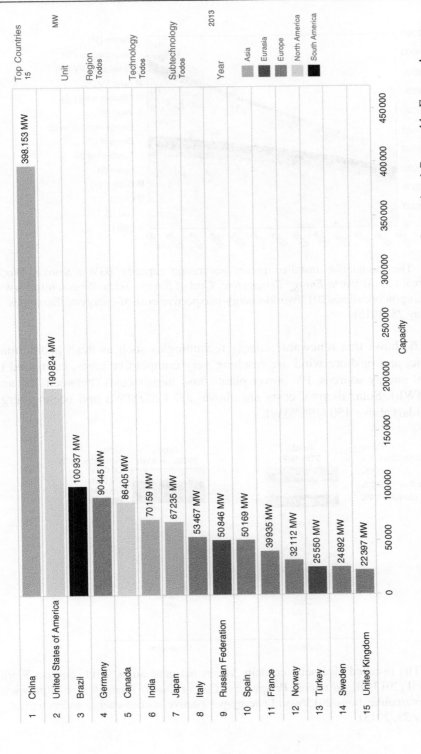

Figure 2.3 The ranking of the renewable power capacity for various countries. *Source*: International Renewable Energy Agency (IRENA). Reproduced with permission of International Renewable Energy Agency (IRENA).

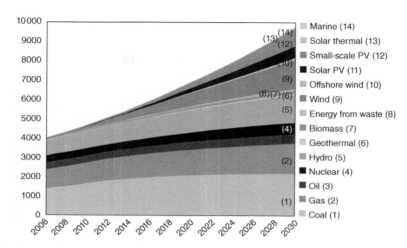

Figure 2.4 The cumulative installed power generation capacity (GW). *Source*: World Energy Council (2013) *World Energy Perspective: Cost of Energy Technologies*, http://www. worldenergy.org/publications/2013/world-energy-perspective-cost-of-energy-technologies/ (accessed May 28, 2015).

Figure 2.6 shows that renewable energy technologies such as flash geothermal, hydroelectric, and onshore wind are reaching very competitive costs, compared to conventional energy sources. PV power plants have higher costs (between 100 and 150 US$/MWh). Solar thermal costs are above 200 US$/MWh and ocean energy (wave and tidal) above 450 US$/MWh.

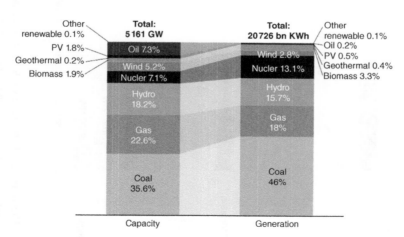

Figure 2.5 The installed electricity capacity versus net generation, 2011. *Source*: World Energy Council (2013) *World Energy Perspective: Cost of Energy Technologies*, http://www. worldenergy.org/publications/2013/world-energy-perspective-cost-of-energy-technologies/ (accessed May 28, 2015).

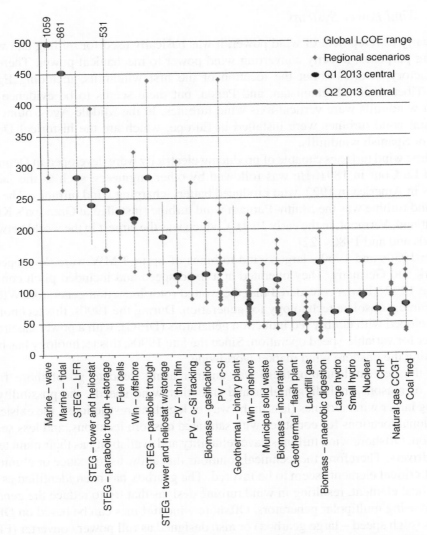

Figure 2.6 The globalized LCOE for Q2 2013 (US$/MWh). *Source*: World Energy Council (2013) *World Energy Perspective: Cost of Energy Technologies*, http://www.worldenergy.org/publications/2013/world-energy-perspective-cost-of-energy-technologies/ (accessed May 28, 2015).

While hydropower can obviously be considered the main renewable energy source, the present book is concentrated on PV solar and wind energy, which have some specific characteristics that make them of greater interest: their instantaneous available power cannot be controlled (although it can be forecasted), they are typically connected to the grid using power electronic converters, and they are dominating worldwide in terms of new installed power plants.

2.1.1 Wind Power Systems

In the early development of wind power, it was basically used for milling and water pumping purposes, directly converting wind power to mechanical power. There are contradictory theories about the location of the first windmills, including Egypt, China, Tibet, India, Afghanistan, and Persia, but there seems to be evidence that the first windmills were vertical-axis wind turbines. In the Middle Ages, numerous horizontal wind turbines were installed in Europe, which are the historical Dutch, Greek, or Spanish windmills.

The first wind turbines capable of producing electricity were developed in Denmark by Paul La Cour in 1891. He was followed by other engineers, such as the Jacobs brothers in America in 1922, who produced battery-charging wind turbines. The first MW wind turbine was the Smith–Putnam wind turbine (installed at Grandpa's Knob, near Rutland, Vermont, in the early 1940s), which was the largest in the world between the 1940s and and 1980s [22].

In the 1950s and 1960s, larger wind turbines of up to 3 MW were developed in Denmark and Germany. They operated at a fixed speed and included pitch control. Until the 1990s, it was considered that the wound rotor induction generator (WRIG) was an almost obsolete technology for generation. During the 1990s, this technology was developed with doubly fed induction generators (DFIG), with a power electronic converter for variable-speed operation. Since the late 1990s, this technology has been dominating the onshore wind market worldwide.

With the new millennium, there was an increasing trend to go offshore for a number of reasons, including higher and smoother wind speeds, the possibility of installing larger wind turbines (less in the way of logistical restrictions), the existence of additional locations for countries with saturated onshore locations, and less social opposition. Offshore wind turbines have reliability and availability as their main technology drivers. Therefore, the technical solutions that allow us to reduce or eliminate the most critical elements seem to be favored. The gearbox has been identified as the most critical element, resulting in wind turbine designs that try to reduce the generator speed using multipolar generators. Offshore wind turbines can be based on DFIG concepts (high speed – large gearbox) or also designed as full power converter (FPC) wind turbines, connected to an induction (high speed – large gearbox) or permanent magnet (low or medium speed, reduced gearbox, multipolar generator) synchronous generator.

While, some decades ago, wind was seen as an alternative energy source, nowadays it can be considered as a mature technology that is being developed massively. According to the Global Wind Energy Council [23], at the end of 2013, the total amount of installed wind power worldwide exceeded 318 GW (see Figure 2.7). During 2013 alone, more than 35 GW of wind power was installed. Based on the information of the Global Wind Energy Council [23], Figure 2.8 shows the cumulative and new installed wind power.

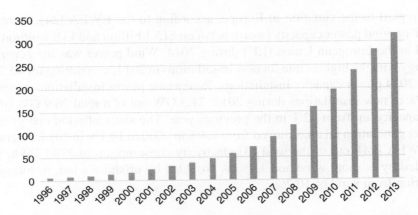

Figure 2.7 The evolution of cumulative installed wind power from 1996 to 2013. *Source*: Adapted from GWEC.

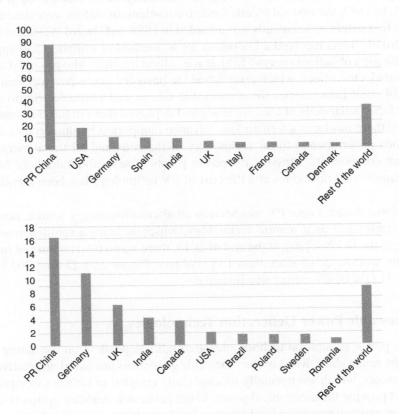

Figure 2.8 The cumulative (top) and new (bottom) installed wind power, December 2013. *Source*: Adapted from GWEC.

With regard to the situation in Europe, according to the EWEA [24], more than 11.5 GW of wind power capacity (worth between €13.1 billion and €18.7 billion) was installed in the European Union (EU) during 2014. Wind power was the generating technology with the highest rate for new installations in 2014, accounting for 43.7% of the total 2014 power capacity installations. Renewable power installations accounted for 79.1% of new installations during 2014: 21.3 GW out of a total 26.9 GW of new power capacity, up from 72% in the previous year. The share of wind energy in the EU power generation mix has moved from 2.4% in 2000 to 14.1% in 2014. According to the EWEA [24], out of the total EU electricity consumption of 2798 TWh, 9.1% is provided by onshore wind and 1.1% is provided by offshore wind, resulting in a contribution from wind of 10.2%.

2.1.2   Solar Photovoltaic Power Systems

The basic principle of the conversion of light to electricity was first observed in 1839 by Becquerel. In 1883, the first solar cells made from selenium wafers were described by Fritts. The first patent on solar cells was awarded in 1888, to Edward Weston in the US. In 1901, Nikola Tesla received a US patent for a "method of utilizing, and apparatus for the utilization of radiant energy" [25]. It was Albert Einstein who provided a theory on the photoelectric effect, which established the basis for future development. In the US in 1954, Bell Laboratories discovered that silicon had photoelectric properties. They developed solar cells of 6% efficiency and deployed them in the early satellites. Since then, there has been a tremendous advance (supported by numerous subsidies from various governments) in the technology of the cells and all the associated PV power plant components. Massive PV power plants of several hundreds of MW are being commissioned nowadays and the cost of PV technology has been significantly reduced.

While, some decades ago, PV was seen as an alternative energy source, nowadays it can be considered as a mature technology, which is being exploited massively. According to the IRENA [20], at the end of 2013, there was a total amount of installed PV capacity worldwide of more than 135 GW (see Figure 2.9). During 2013 alone, more than 37 GW of PV power was installed.

2.2   Renewable Power Generation Technologies

Renewable power generation technologies are highly dependent on the nature of the source of the renewables. Most of the renewable generators are based on rotative electrical generators, which are normally mechanically coupled to turbines or equivalent devices that provide the mechanical power to the generator. Another option is to have static power generation sources, which can produce electricity without any moving parts, as in the case of photovoltaic generation systems. The various technologies are examined in this section.

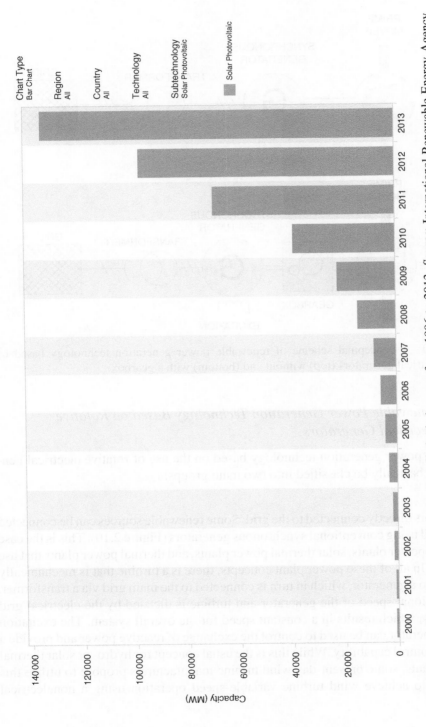

Figure 2.9 The evolution of cumulative installed wind power from 1996 to 2013. *Source*: International Renewable Energy Agency (IRENA). Reproduced with permission of International Renewable Energy Agency (IRENA).

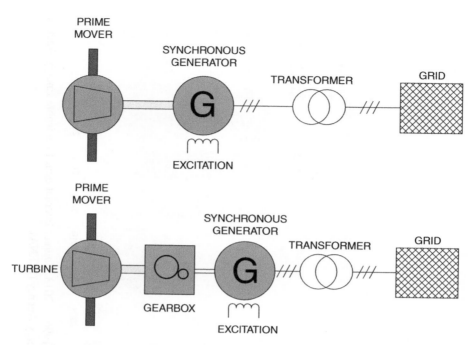

Figure 2.10   A conceptual scheme of renewable power generation technology based on rotative electrical generators (top) without and (bottom) with a gearbox.

2.2.1   Renewable Power Generation Technology Based on Rotative Electrical Generators

Renewable power generation technology based on the use of rotative electrical generators can basically be classified into two main groups:

- Generators directly connected to the grid. Some renewable sources can be connected to the grid using conventional synchronous generators (Figure 2.10). This is the case for hydropower plants, solar thermal power plants, and thermal power plants that use biofuels. In all of these power plant concepts, there is a turbine that is mechanically coupled to a generator, which in turn is connected to the main grid via a transformer. The rotational speed of the generator and turbine is dictated by the electrical grid frequency, which results in a constant speed for the overall system. The excitation of the generator can be used to control the exchange of reactive power and provide a voltage control capability. While this is the usual concept for hydro- or solar thermal power plants, some present-day wind turbine manufacturers propose to utilize this concept, to achieve wind turbine variable-speed operation using a nonelectrical principle.

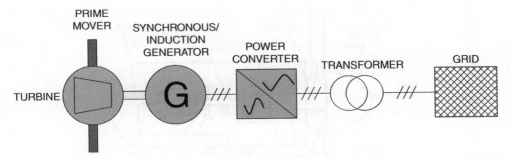

Figure 2.11 The conceptual scheme of renewable power generation technology based on rotative electrical generators with power electronics converters.

- Generators connected to the grid by means of power electronic converters (Figure 2.11). Power electronic converters allow us to control the voltage, current, and frequency of the generator, thus allowing the renewable energy generators to operate at variable speed, while providing support to the grid where they are connected. Most wind turbines are based on this technology, which is also preferred for ocean energy sources (wave or tidal). Power converters can decouple the generator from the grid, which can be very advantageous in terms of fault behavior, but this is also accompanied by additional costs and concerns about reliability.

  The present section focuses on wind turbine technology, since this is considered the most relevant technology for the development of renewable energy using devices equipped with a rotating generator. Similar technologies can be utilized by other RES based on nondispatchable fluid power sources, including ocean energy (waves and tidal) or micro- or mini-hydropower plants.

2.2.2 Wind Turbine Technology

Various wind turbine concepts have been developed in recent decades. Wind turbines can classified into those that feature a vertical axis, such as the Savonius, Darrieus, or Giromill rotors, or horizontal-axis turbines. Nowadays, the dominant concept is the well-known horizontal-axis turbine with a three-bladed rotor.

Figures 2.12 and 2.13 show conceptual schemes of onshore and offshore wind turbines. The rotor is the main rotating element, and it is composed of the hub, the blades, and the mechanical shaft. The blades capture energy from the wind, and transfer it to the generator by means of the shaft. The pitch system allows us to limit the incoming power from the wind by changing the blade angle, in order to keep the machine within its power limits. Usually, wind turbines include a gearbox in order to adapt the slow angular speed of the rotor blades to the high speed of the generator. This gearbox can also be removed in some topologies, if the generator is modified

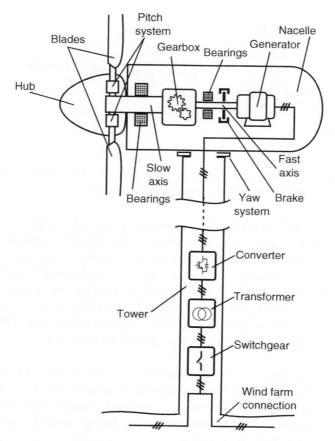

Figure 2.12 The conceptual scheme of the horizontal-axis onshore wind turbine.

so that its rotor speed can be adapted to the speed of the electrical generator; in, for instance, multipolar schemes.

The high-speed gearbox output shaft (the fast axis) is directly connected to the generator, where the mechanical-to-electrical power conversion is performed. The generator is an induction or synchronous machine. Depending on the generator installed and the turbine concept considered, power electronics converters could be connected at the machine rotor or stator terminals. By means of this converter, the generator can be controlled, achieving better wind power extraction, together with proper integration into the grid. The generator output is connected to a transformer in order to adapt the voltage level and its corresponding switchgear is placed ahead of the connection to the wind power plant collection grid.

The previous component description applies to both onshore and offshore wind turbines. The main difference between them is the support structure employed. Onshore

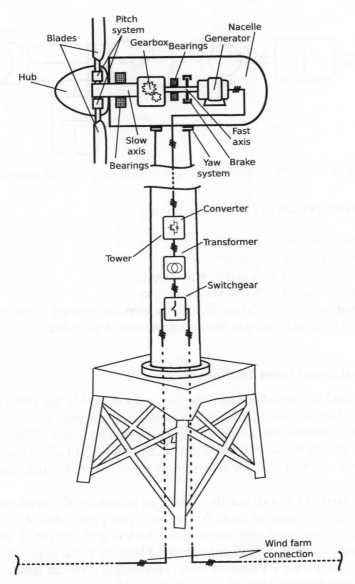

Figure 2.13 The conceptual scheme of the horizontal-axis offshore wind turbine.

wind turbines are installed over a conventional concrete basement, whereas offshore wind turbines demonstrate different possibilities. Two main offshore structures can be differentiated, the foundation and floating types. Foundation structures are based on supporting the wind turbine by means of a structure attached to the seabed. Floating structures employ buoyancy or ballast techniques.

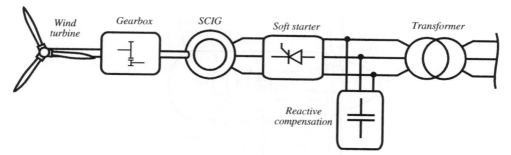

Figure 2.14   The SCIG fixed-speed wind turbine, with a multiple-stage gearbox.

Wind turbine types can be classified as follows:

- Type 1. Fixed-speed wind turbine
- Type 2. Limited-speed wind turbine
- Type 3. Variable-speed with partial-scale converter
- Type 4. Variable-speed with full-scale converter
- Type 5. Variable-speed with a nonelectrical conversion principle – generator directly connected to the grid using synchronous generators with excitation.

2.2.2.1   Fixed-Speed Wind Turbines

Fixed-speed wind turbines dominated the European market in the 1980s and 1990s. Known as "the Danish concept," these devices are usually feature a three-bladed rotor, a multiple-stage gearbox, and a squirrel-cage induction generator (SCIG), connected directly to the grid by means of a transformer [26] (Figure 2.14). This machine topology is operated over and near the synchronous speed, where the induction machine works as a generator.

The SCIG needs to absorb reactive power to magnetize the machine, which is usually provided by a capacitor bank. With regard to the power control, these turbines usually include a thyristor-based soft-starter, which is used to start up the machine when it is connected to the grid in order to avoid large currents. A possible technical solution is to incorporate a generator with two possible different numbers of poles to modify the synchronous speed, increasing the wind power extraction possibilities. These machines can be divided into three different types, depending on the power limiting system implemented; that is, passive or active stall, and pitch [27].

The main advantages of fixed-speed generation are the robustness of the system, the relatively low production costs, and its straightforward control [28]. These advantages were the main reasons that prompted the manufacturers to install this technology in the first wind farms. However, it also presented several important disadvantages that have forced the change to other topologies. The first drawback is related to

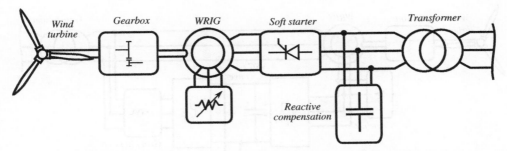

Figure 2.15 The SCIG limited variable-speed turbine, with a multiple-stage gearbox.

the constant-speed operation of the generator, which does not allow the machine to extract the maximum available power from the wind. Also, a fixed speed implies high mechanical stress, considering that any wind speed fluctuations are transmitted in terms of torque, causing possible failures of the drivetrain. Other drawbacks were the requisites imposed by the grid operator, such as voltage/frequency support and the fault ride-through requirements. Not only cannot the SCIG provide voltage support by injecting reactive power, but it also needs to absorb it during its normal operation. Regarding the fault ride-through capability, SCIG wind turbines are connected directly to the grid and it is difficult to avoid a disconnection due to a high current flowing through the generator during a voltage sag [28].

2.2.2.2 The Limited Variable-Speed Wind Turbine

During the 1990s, the limited variable-speed turbine was proposed, changing the generator type from the conventional SCIG to a WRIG [29]. This technology (Figure 2.15), also known under the commercial name *Optislip* (Vestas manufacturing), introduced a variable rotor resistance to increase the operational speed range (typically from 0 to 10%). By changing the value of the rotor resistance, through a power electronic converter, the slip (and speed) of the machine can be modified. However, in order to achieve this level of control, some power has to be lost in the wound rotor resistances. In summary, these machines partially solved the demand of the need for a variable speed to increase the aerodynamic efficiency, but they still mainly suffered from the same problems as the fixed-speed concept. This was one of the first steps towards the variable-speed concept, which dominates the wind turbine market today.

2.2.2.3 Variable-Speed Wind Turbines with a Partial-Scale Converter

The variable-speed concept appeared in order to fulfill the grid operator's requirements and also to increase the operational performance of wind turbines. This concept

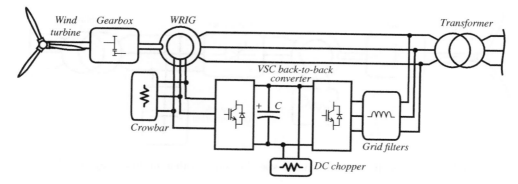

Figure 2.16 The DFIG variable-speed turbine, with a multiple-stage gearbox.

(Figure 2.16) consists of a WRIG generator connected to the grid directly by the stator and through the rotor by a partial-scale power converter, resulting in the so-called doubly fed induction generator (DFIG) [30]. This power converter allows us to control the machine in a wider speed range (typically from −40% to +30%) than the fixed- or limited-speed topologies. Using the converter, proper active and reactive power control can be carried out in order to optimize the wind power extraction of the machine.

With regard to integration into the grid, the partial-scale converter increases the possibilities of the system. The grid-side part, operated independently from the machine side, can be controlled to achieve a better ride-through capability of the wind turbine [31] and for voltage support.

However, this concept suffers from various drawbacks [27]. Slip rings, one of the components that cause operational machine failures, are needed in order to extract the power from the rotor. Moreover, the stator is still connected directly to the grid, so during grid faults it is complicated to manage a proper ride-through operation and to avoid mechanical overloads in the gearbox during electrical disturbances.

2.2.2.4 Full-Scale Converter Wind Turbines

Variable-speed wind turbines with full-scale converters electrically decouple the generator and the grid completely. Additionally, they also maximize the operational speed range of the wind turbine. Therefore, the maximum available power from the wind can be extracted within the rated limits. The grid operator's requirements can be satisfied, with proper grid-side converter control. These systems can give support to the grid voltage and frequency and they can incorporate fault ride-through capability. Usually, a DC chopper is included in the DC bus to ride through voltage sags without changing the mechanical torque of the turbine, thus avoiding mechanical overloads in the wind turbine. This concept can be implemented for different types of generator: the

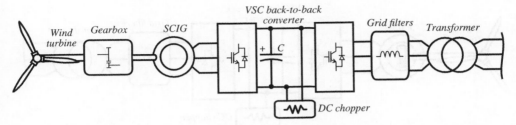

Figure 2.17   The SCIG variable-speed turbine, with a multiple-stage gearbox.

SCIG, the permanent magnet synchronous generator (PMSG), and the wound rotor synchronous generator (WRSG) [32].

The SCIG
The drivetrain is composed of a SCIG and a full power converter [33] (Figure 2.17). This concept is not available without a gearbox, due to the inconvenience in construction of having multipolar SCIGs [34]. The gearbox raises concerns for offshore applications, since it may compromise the wind turbine's reliability, availability, and maintainability, which are very critical offshore. Another drawback is the relatively low efficiency of the global system, considering the generator, gearbox, and full power converter losses [26].

The Direct-Drive WRSG
The drivetrain is composed of a WRSG with DC rotor excitation connected to the grid by means of a back-to-back converter (Figure 2.18). This type is implemented without a gearbox through incrementing the number of poles of the rotor, thus obtaining a directly driven system. The increment on the number of poles implies a large generator diameter, resulting in a machine of larger dimensions [35]. On the one hand, the associated costs of the gearbox are eliminated. On the other hand, the generator costs are increased. Another interesting feature of this machine is the DC controllable

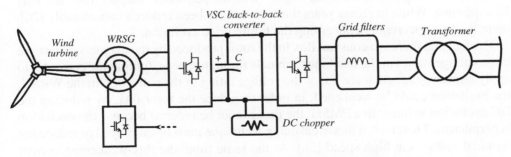

Figure 2.18   The WRSG variable-speed turbine, with direct drive.

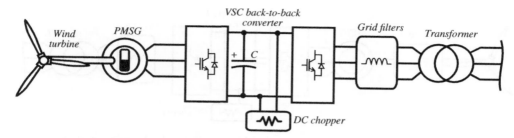

Figure 2.19   The PMSG variable-speed wind turbine, with direct drive.

excitation. Through the DC converter, the rotor flux can be controlled for different purposes: field-weakening strategies for high-speed operation and loss minimization techniques, amongst others. However, it must be taken into consideration that the excitation control is performed by a converter, which is located outside the rotor and is commonly connected to the machine windings through slip rings. This system implies losses during operation due to the current flowing the DC windings, and also regular maintenance must be considered if slip rings are used.

The Direct-Drive PMSG

Related to the previous WRSG concept, the PMSG direct-driven generator connected to the grid by means of a full power converter may be used (Figure 2.19). This type of generator is derived from the idea of substituting the wound-rotor DC excitation by magnets, thus eliminating the excitation losses.

The PMSG concept, like all the concepts based on full-scale converters, allows decoupling of the generator from the grid. PMSG rotors can also be constructed with a large number of poles. However, PMSGs are smaller and lighter when a large number of poles are included than the equivalent WRSG [35]; therefore, this topology achieves a higher power density.

On the other hand, working with permanent magnets can be difficult during both the manufacturing and operational stages. Also, the permanent magnet materials may be expensive. While in recent years these costs have been reduced considerably [26], there is some uncertainty concerning the future price evolution.

The fact that there is a constant flux in the rotor produced by the magnets may cause large voltage values at the machine terminals for high speeds, forcing the converter to apply larger voltages that are out of the voltage rating of the converter. In the WRSG, the excitation could be weakened, in order to reduce the rotor flux, by reducing the DC excitation voltage. In a PMSG, the flux cannot be reduced because the excitation is permanent. Therefore, a flux-weakening technique must be employed to reduce the terminal voltages at high speed [36]. At the same time, the flux-weakening current

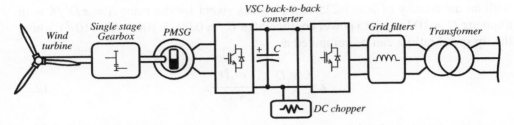

Figure 2.20   The PMSG variable-speed wind turbine, with a single-stage gearbox.

must be limited in order not to demagnetize the magnets. Demagnetization can be caused by certain possible faults in the machine. Magnets are temperature-sensitive, and therefore the temperature should be controlled during operation. There are many different types of PMSG, which can be classified by their flux direction, the type of rotor, the type of the magnets, and the winding layout.

The PMSG Wind Turbine with a Single-Stage Gearbox and a Medium-Speed Generator

This concept combines a PMSG with a single-stage or a multistage gearbox system in order to avoid constructing a large generator with a large number of poles [35] (Figure 2.20). The single-stage topology does not achieve a high transmission ratio. It is interesting because it combines a small gearbox (compared to a fast speed generator, SCIG) with a relatively small generator with a small number of poles (compared to multipolar concepts). Therefore, the generator does not need to be too large and expensive, and the complexity of the gearbox can also be reduced. The generator is operated with medium speed and torque characteristics.

2.2.2.5   Example 1: A Power Computation for Fixed-Speed Wind Turbines

A 3.2 MW wind turbine with a rotor diameter of $D = 100$ m and a gearbox ratio of $N_{gb} = 80$ based on an induction generator (with two pole pairs) is connected to a 960 V 50 Hz grid. An air density of $\rho = 1.225$ kg/m^3 and a power coefficient expression of $C_p = 0.0045(100 - (\lambda - 10)^2)$ can be assumed. Neglecting slip (assuming that $s = 0$), calculate the mechanical power generated for wind speeds of 5, 8, 11, and 14 m/s.

The mechanical power generated can be found using

$$P_{mech} = \frac{1}{2}\rho C_p A v_w^3, \tag{2.1}$$

with an air density of $\rho = 1.225$ kg/m^3, areas swept by the rotor $A = \pi D^2/4$ with diameter $D = 100$ m, and a power coefficient of $C_p = 0.0045(100 - (\lambda - 10)^2)$, where the tip speed ratio λ can be computed as

$$\lambda = \frac{\omega_t D}{2v_w} = \frac{\omega_g D}{2N_{gb}v_w}, \tag{2.2}$$

in which ω_t is the wind turbine speed (slow axis), ω_g is the generator speed (fast axis), and N_{gb} is the gearbox ratio.

The previous expressions can be combined as

$$P_{mech} = \frac{0.0045}{2}\rho(100 - (\lambda - 10)^2)Av_w^3, \tag{2.3}$$

$$P_{mech} = \frac{0.0045}{2}\rho\left(100 - \left(\frac{\omega_g D}{2N_{gb}v_w} - 10\right)^2\right)v_w^3\pi D^2/4, \tag{2.4}$$

where all the quantities are known except for the generator speed ω_g. The generator speed depends mainly on the slip s and the synchronous speed ω_s:

$$\omega_g = (1 - s)\omega_s = (1 - s)\frac{2\pi f}{p}, \tag{2.5}$$

where the p are pairs of poles. Assuming $s = 0$,

$$\omega_g = \omega_s = 2\pi 50/p = \pi 50 \, \text{rad/s}. \tag{2.6}$$

Therefore, the mechanical power can be computed as follows:

$$P_{mech} = \frac{0.0045}{2}\rho\left(100 - \left(\frac{\omega_s D}{2N_{gb}v_w} - 10\right)^2\right)v_w^3\pi D^2/4, \tag{2.7}$$

$$P_{mech} = \frac{0.0045}{2}1.225\left(100 - \left(\frac{\pi 50 \times 100}{2 \times 80 \times v_w} - 10\right)^2\right)v_w^3\pi 100^2/4, \tag{2.8}$$

The following results are obtained:

- $\omega_e = 3.141592653589793e+002$ rad/s
- $\omega_s = 1.570796326794897e+002$ rad/s
- $\omega_g = 1.570796326794897e+002$ rad/s
- $\omega_t = 1.963495408493621$ rad/s

- $\lambda = 19.634954084936204$
- $C_p = 0.032254469015270$
- $P = 1.939527243080785e+004$ W.

The calculations can be extended to the other wind speeds, obtaining:

$v_w = 5$	8	11	14	m/s
$\omega_s = 314.1593$	314.1593	314.1593	314.1593	rad/s
$\omega_g = 157.0796$	157.0796	157.0796	157.0796	rad/s
$\omega_t = 1.9635$	1.9635	1.9635	1.9635	rad/s
$\lambda = 19.6350$	12.2718	8.9250	7.0125	–
$C_p = 0.0323$	0.4268	0.4448	0.4098	–
$P = 1.9395e+004$	1.0511e+006	2.8480e+006	$P = 3.2$ **MW**	W

where it can be noted that when the mechanical power exceeds the nominal power (3.2 MW), the pitch system must reduce the mechanical power to its nominal value.

The MATLAB® code that can be used for the analysis

```
clear;clc;
vw=[5 8 11 14]
ss=0;rho=1.225;N=80;D=100;poles=2;Ugrid=960;f=50;
we=2*pi*f;
ws=we/poles;
wg=ws*(1-ss);
wt=wg/N;
lam= wt*(D/2) ./ vw
Cpp= 0.0045 * (100 - (lam-10).^2)
P1=0.5*rho*Cpp*(pi*(D^2)/4).*vw.^3
```

results in

```
vw =     5     8    11    14
lam =    19.634954084936204   12.271846303085129
8.924979129516457   7.012483601762931
Cpp =    0.032254469015270   0.426774214688213
0.444799485576112   0.409836355966191
P1 =    1.0e+006 *
0.019395272430808   1.051148578283138
2.847988990930852   5.409916510373950
```

The C_p–λ curve can be plotted using the following code (the results are shown in Figure 2.21):

```
%% Develop a Matlab program to locate the operating
points in the Cp-lambda curve.
```

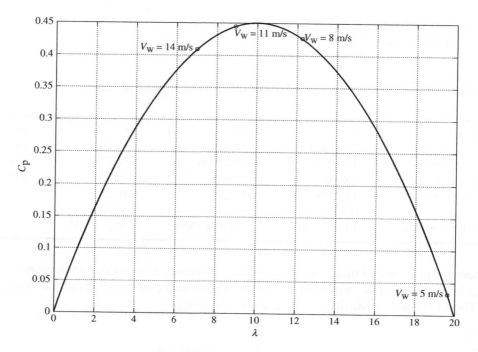

Figure 2.21 The $C_p-\lambda$ curve.

```
lam1=0:.1:20;
Cpp1= 0.0045 * (100 - (lam1-10).^2);
plot(lam1,Cpp1);hold on;grid on;
plot(lam,Cpp,'ko');
for ii=1:1:4
txt{ii}=['v_w = ' num2str(vw(ii)) ' m/s'];
text(lam(ii),Cpp(ii),txt{ii},'FontSize',18);
end;
xlabel('\lambda','FontSize',18);
ylabel('C_p','FontSize',18);
```

The power curves can be analyzed as follows (the results are shown in Figure 2.22):

```
%% Develop a Matlab program to plot the power
%% generated for different wind  speeds. Include
%% a comparison between the obtained power and
%% the maximum available power.

figure(2);
vw2=0:.1:15
lam2= min(20,max(wt*(D/2) ./ vw2,0));
```

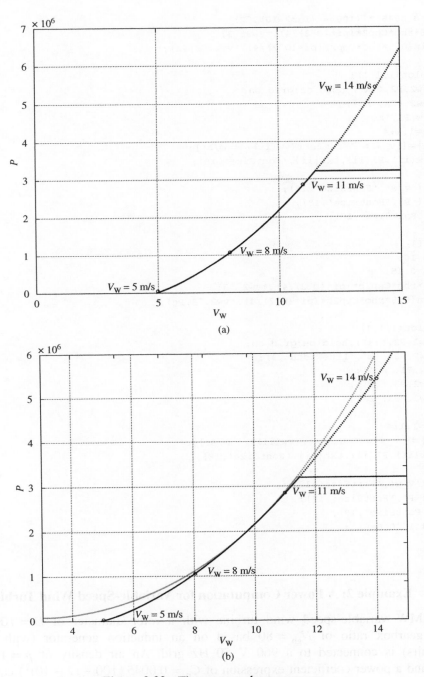

Figure 2.22 The generated power curve.

```
Cpp2= 0.0045 * (100 - (lam2-10).^2)
P2=0.5*rho*Cpp2*(pi*(D^2)/4).*vw2.^3;
P2a=min(0.5*rho*Cpp2*(pi*(D^2)/4).*vw2.^3,Pn);

h=subplot(1,1,1);
plot(vw2,P2,':k');hold on;grid on;
plot(vw2,P2a,'k');
plot(vw,P1,'ko');
for ii=1:1:4
txt{ii}=['v_w = ' num2str(vw(ii)) ' m/s'];
text(vw(ii),P1(ii),txt{ii},'FontSize',18);
end;
xlabel('v_w','FontSize',18);
ylabel('P','FontSize',18);
set(h,'FontSize',18);

figure(3);
lam2b= 10;
Cpp2b= 0.45;
P3=0.5*rho*Cpp2b*(pi*(D^2)/4).*vw2.^3;
P3a=min(0.5*rho*Cpp2b*(pi*(D^2)/4).*vw2.^3,Pn);

h=subplot(1,1,1);
plot(vw2,P2,':k');hold on;grid on;
plot(vw2,P2a,'k','LineWidth',3);
plot(vw,P1,'ko');
plot(vw2,P3,':r');
plot(vw2,P3a,'r');

for ii=1:1:4
txt{ii}=['v_w = ' num2str(vw(ii)) ' m/s'];
text(vw(ii),P1(ii),txt{ii},'FontSize',18);
end;
xlabel('v_w','FontSize',18);
ylabel('P','FontSize',18);
set(h,'FontSize',18);
axis([3 15 0 6e6]);
```

2.2.2.6 Example 2: A Power Computation for Variable-Speed Wind Turbines

A 3.2 MW variable-speed wind turbine with a rotor diameter of $D = 100$ m and a gearbox ratio of $N_{gb} = 80$ based on an induction generator (with two pole pairs) is connected to a 960 V 50 Hz grid. An air density of $\rho = 1.225$ kg/m^3 and a power coefficient expression of $C_p = 0.0045 \left(100 - (\lambda - 10)^2\right)$ can be assumed. Calculate the mechanical power generated for wind speeds of 5, 8, and 11 m/s.

The mechanical power generated can be found using

$$P_{mech} = \frac{1}{2}\rho C_p A v_w^3,$$ (2.9)

with an air density of $\rho = 1.225$ kg/m^3, areas swept by the rotor $A = \pi D^2/4$ with diameter $D = 100$ m, and a power coefficient of $C_p = 0.0045(100 - (\lambda - 10)^2)$, where the tip speed ratio λ can be computed as

$$\lambda = \frac{\omega_t D}{2v_w} = \frac{\omega_g D}{2N_{gb}v_w},$$ (2.10)

in which ω_t is the wind turbine speed (slow axis), ω_g is the generator speed (fast axis), and N_{gb} is the gearbox ratio.

The previous expressions can be combined as follows:

$$P_{mech} = \frac{0.0045}{2}\rho(100 - (\lambda - 10)^2)A v_w^3,$$ (2.11)

$$P_{mech} = \frac{0.0045}{2}\rho \left(100 - \left(\frac{\omega_g D}{2N_{gb}v_w} - 10\right)^2\right) v_w^3 \pi D^2/4,$$ (2.12)

where all the quantities are known except for the generator speed ω_g.

The main difference compared with the previous example is that in the present exercise a variable-speed wind turbine is considered. Therefore, the speed of the wind turbine depends on the optimal operation point and not on the electrical grid frequency. The optimal operation point can be found by differentiating the expression for C_p:

$$\frac{dC_p}{d\lambda} = \frac{d\left[0.0045\left(100 - (\lambda - 10)^2\right)\right]}{d\lambda} = -0.0045 \times 2(\lambda - 10) = -0.009(\lambda - 10),$$ (2.13)

$$\frac{dC_p}{d\lambda} = 0 \rightarrow \lambda_{opt} = 10,$$ (2.14)

$$\frac{d^2 C_p}{d\lambda^2} = -0.009 < 0 \rightarrow \lambda_{opt} = 10 \rightarrow \text{maximum}.$$ (2.15)

The wind turbine speed can be calculated as

$$\omega_t = \frac{2\lambda_{opt}v_w}{D}$$ (2.16)

and the generator speed as

$$\omega_g = N\omega_t. \tag{2.17}$$

The following results are obtained:

v_w	5	8	11	m/s
λ_{opt}	10	10	10	
C_p	0.45	0.45	0.45	
ω_t	1	1.6	2.2	rad/s
P_{mech}	0.27059	1.1084	2.8813	MW

The wind speed that reaches nominal power (assuming two operational regions, partial power and maximum power) can be calculated as

$$P_{mech-nom} = \frac{1}{2}\rho C_{p-max} A v_w^3 \rightarrow v_w = 3\sqrt{\frac{2P_{mech-nom}}{\rho C_{p-max}A}} = 11.391 \text{ m/s}. \tag{2.18}$$

Above this wind speed, how is the wind turbine operated? Calculate the turbine speed, the power coefficient, the tip speed ratio, and the mechanical power generated for a wind speed of 14 m/s.

The wind turbine speed is maintained constant at the nominal value and the pitch system is used to maintain the power at the nominal value. It can be assumed that the nominal speed is the turbine speed for the wind speed of the previous section. Therefore

$$\omega_t = \frac{2\lambda_{opt}v_w}{D} = 2.278297808332945 \text{ rad/s}. \tag{2.19}$$

For a wind speed above 11.39 m/s, the mechanical power will be the nominal power of 3.2 MW. The C_p for a wind speed of 14 m/s can be calculated from

$$P_{mech} = \frac{1}{2}\rho C_p A v_w^3 \rightarrow C_p = \frac{2P_{mech}}{\rho A v_w^3} = 0.242420809374220. \tag{2.20}$$

The tip speed ratio cannot be obtained from the expression for C_p, since this expression does not take the pitch angle into consideration. As the wind turbine rotational speed has been previously calculated, the tip speed ratio can be also calculated as

$$\lambda = \frac{\omega_t D}{2v_w} = 8.136777886903376. \tag{2.21}$$

2.2.3   Photovoltaic Power Plants

PV power plants can be considered as renewable power generation technology based on static devices. There a number of different PV generation plants that can be considered and they are of very different natures. PV can be used in domestic applications, in both grid-connected and off-grid configurations. In the off-grid case, the PV generator plant will need to be complemented by other renewable generators, energy storage devices, and/or backup diesel generators. Also, large PV power plants are being designed, covering several km^2 of ground surface and reaching a peak power of several hundred MW. As can be seen in Figure 2.23, a PV power plant is based on various strings, trackers, or groups of PV panels, which are connected to a power converter to inject the power into the AC collection grid. A substation collects all the power from the different inverters and injects it into the main grid. There are several possible combinations and arrangements for a PV power plant, including DC collection grids and the utilization of smaller inverters. Figures 2.24, 2.25, and 2.26 show PV power plants based on central, string, or module inverters, respectively. The decision regarding the configuration depends on the total cost of the plant and also on the expected losses associated with loss of power due to eventual shadows, and availability and reliability issues.

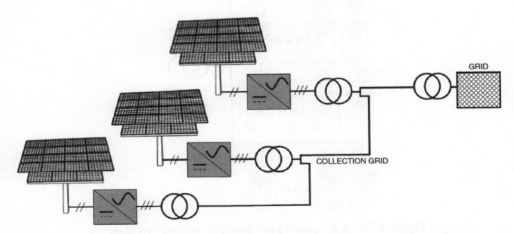

Figure 2.23   The conceptual scheme of a PV renewable generation system.

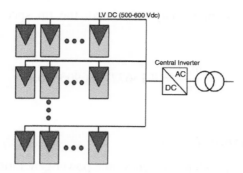

Figure 2.24 A PV power plant based on a central inverter.

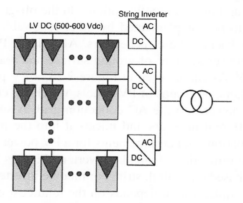

Figure 2.25 A PV power plant based on string inverters.

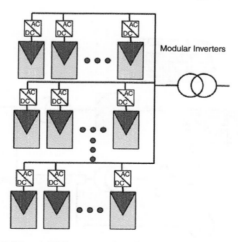

Figure 2.26 A PV power plant based on module inverters.

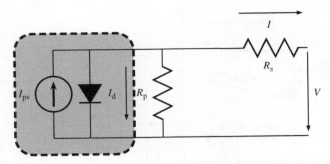

Figure 2.27 The PV panel model.

The energy capture basically depends on the radiation and the temperature (as discussed in the following example). In order to maximize the energy capture, PV systems can be equipped with trackers that orientate the panels to the sun. The trackers can allow rotation around one or two axes. If no trackers are used, the panels will be oriented depending on the application. For grid-connected systems, the panels will be oriented in order to maximize the annual energy production. For off-grid systems, the panels will be oriented so as to ensure demand provision in the season with the lowest energy production.

2.2.3.1 Example 3: The Analysis of Photovoltaic Panels

Using the model of Walker [37], an example of a PV model from Solarex can be analyzed and the V–I and P–I characteristics can be obtained. The model used is sketched in Figure 2.27. The various currents can be calculated as

$$I = I_{\mathrm{PV}} - V_0 \left[\exp\left(\frac{V + R_s I}{V_t a}\right) - 1 \right] - \frac{V + R_s I}{R_p},$$

$$I_{\mathrm{PV}} = \left(I_{\mathrm{PV},n} + K_{\mathrm{I}}\left(T - T_n\right)\right) \frac{G}{G_n},$$

$$I_0 = \frac{I_{sc,n} + K_{\mathrm{I}}\left(T - T_n\right)}{\exp\left(\dfrac{V_{oc,n} + K_{\mathrm{V}}\left(T - T_n\right)}{a V_t}\right) - 1},$$

(2.22)

where the shunt and series resistances can be calculated as

$$R_{s,n} = \frac{N_s}{N_p} R_s, \quad R_{p,n} = \frac{N_s}{N_p} R_p.$$

(2.23)

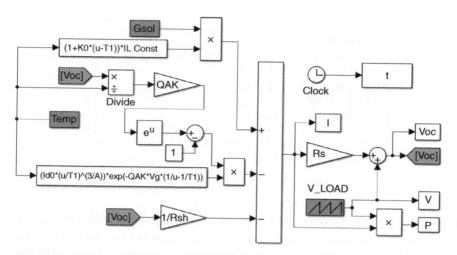

Figure 2.28 The simulation model employed for the analysis of PV modules.

The following MATLAB® script introduces all of the parameters and makes some initial precalculations for later use in the Simulink model sketched in Figure 2.28. The results obtained are shown in Figure 2.29.

```
Temp=298; %Temperature K
Gsol=1000; % Radiation W/m2
Rs=0.008; % Rseries
Rsh=200; % Rshunt
A=1; % Form factor
Kboltz=1.38e-23; % Boltzman constant
Qcharge=1.6e-19; % Electron charge
QAK=Qcharge/(A*Kboltz); % Constant including Q, A and K

% Solarex module data from Walker
T1 = 273 + 25;
Voc_T1 = 21.06 /36;
Isc_T1 = 3.80;
T2 = 273 + 75;
Voc_T2 = 17.05 /36;
Isc_T2 = 3.92;

Id0=Isc_T1/(exp(QAK*Voc_T1/T1)-1);
K0=(Isc_T2-Isc_T1)/(T2-T1);
Vg=1.12; % depn del material
ILConst= 3.8e-3; % Constant I/G -   [A/(Wm-2)]

% Run the simulink model for 3 different
% % temperatures or radiations
```

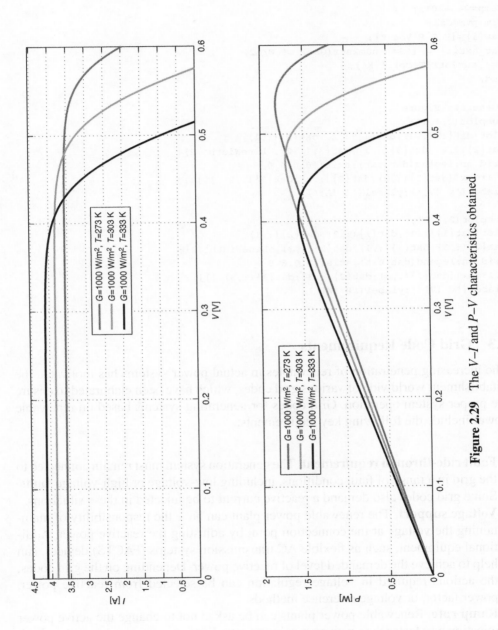

Figure 2.29 The V–I and P–V characteristics obtained.

```
for i=1:1:3;
%Gsol=100+300*i;
Temp=243+30*i;
sim pvmodel;
dat{i}=[V I P Voc t];
dat_txt{i} = ['G=' num2str(Gsol) ' W/m2,
T=' num2str(Temp) ' K'];
end;

%Generate Figure
subplot(2,1,1);
plot(dat{1}(:,1),dat{1}(:,2),dat{2}(:,1),
dat{2}(:,2),dat{3}(:,1),dat{3}(:,2),'LineWidth',2);
grid on;legend(dat_txt);axis([0 0.6 0
1.1*max([dat{1}(:,2);dat{2}(:,2);dat{3}(:,2) ])]);
xlabel('V [V]');ylabel('I [A]');

subplot(2,1,2);
plot(dat{1}(:,1),dat{1}(:,3),dat{2}(:,1),
dat{2}(:,3),dat{3}(:,1),dat{3}(:,3),'LineWidth',2);
grid on;legend(dat_txt);axis([0 0.6 0
1.1*max([dat{1}(:,3);dat{2}(:,3);dat{3}(:,3) ])]);
xlabel('V [V]');ylabel('P [W]');
```

2.3   Grid Code Requirements

The increasing penetration of renewables in actual power systems has motivated the establishment worldwide of various grid codes, which have been elaborated to ensure the proper system operation. Grid codes for generating systems based on renewable power include the following key requirements:

- **Fault ride-through requirement**. The generation system must remain connected to the grid in a range of fault conditions, including low-voltage or high-voltage faults. Some grid codes also demand a reactive current to be injected in these situations.
- **Voltage support**. The renewable power plant can have the responsibility of maintaining the voltage at the connection point by adjusting the reactive power. Additional equipment, such as flexible AC transmission systems (FACTS) devices, can help to achieve the demanded level of reactive power. Depending on the grid codes, the actions required in voltage regulation can be selected from reactive power, power factor, or voltage reference methods.
- **Ramp rate**. Renewable power plants can be asked not to change the active power injection any faster than a given maximum rate. For some renewable sources, such as PV power plants, energy storage systems (ESSs) may have to be able to fulfill this requirement.

- **Frequency support**. Renewable plants can be ordered to adjust their active power injection depending on the grid frequency in order to provide balancing power to ensure system stability. The usual approach is to define a droop characteristic that links the frequency to the required power injection. Normally, it will be simple to reduce the power injection when there is a frequency increase. For the case of a frequency decrease and the need to inject additional power, various solutions can be adopted, including operation with some level of power reserve and the utilization of ESSs.
- **Other**. Other requirements are being discussed in some countries, including virtual inertia emulation or an oscillation damping capability.

2.4   Conclusions

Nowadays, renewable forms of energy is contributing significantly to the energy mix in some countries worldwide. In the future, a substantial increase is expected to be seen in the penetration of renewable energy systems, which may lead to power systems mainly based on renewable energy. Hydro-, wind, and PV solar power plants are leading in terms of installed capacity and energy production, but there are also other renewables that have a promising future (geothermal and solar thermal) and others that are in earlier stages of research and development. Future power systems that include a high penetration of renewables will face some important challenges related to the uncontrollability of some of the main renewable energy sources. One of the main options for tackling this challenge is the utilization of energy storage technologies and active demand-side management. Furthermore, renewable power plants will have to provide ancillary services in order to ensure overall system stability.

Electric power systems are being radically transformed in order to be capable of absorbing all the renewables, while ensuring reliable power system operation. Modern power systems need advanced control and communication systems (the Smart Grid) and also strong actuators (power electronics) to ensure the proper power flows and voltages throughout the system. In this new power system framework, electrical energy storage systems face the challenge of balancing the overall energy flows, allowing electrical grids to be run with affordable costs and enhanced reliability.

3

Frequency Support Grid Code Requirements for Wind Power Plants

This chapter presents the specific case [38] of the participation of wind power plants (WPPs) in system frequency control. While the chapter looks specifically at WPPs, the results can be exported to other renewable energy generation technologies.

For the stable operation of an electrical network, system frequency control is essential. It ensures a continuous adaptation of power generation to power consumption. The power balance in the electrical network is interrelated with the network frequency via all of the synchronous generators connected to it; for example, an increase in the load decelerates the synchronous generators and thus leads to a frequency drop. As frequency is uniform throughout the interconnected network, it is convenient to use it as a control variable for a decentralized control system: the network frequency control. This makes use of the power plants in the network, which – according to their abilities and agreements – adapt their active power feed-in according to the current system requirements. Thus, the power plants involved require a certain level of active power reserves. Traditionally and still typically, it is conventional generation plants, such as hydroelectric and thermal power plants, that are used for frequency control. The ability of a system to maintain its frequency within a certain tolerance band is called the frequency stability.

Another important function of conventional power plants for frequency control is the passively provided so-called *instantaneous power reserve*. Any imbalance between power generation and consumption is instantaneously balanced due to the physical principles of the synchronous generator. The large inertia of the rotating generator set works as buffer storage, with any usage leading to the above-mentioned change in rotational speed and thus in system frequency. The larger the synchronized inertia in the system, the slower is the change of frequency [39].

Energy Storage in Power Systems, First Edition. Francisco Díaz-González, Andreas Sumper and Oriol Gomis-Bellmunt.
© 2016 John Wiley & Sons, Ltd. Published 2016 by John Wiley & Sons, Ltd.

The stepwise replacement of conventional generating units by wind and photo-voltaic power plants will have a significant impact on the behavior of the system frequency. First, the grid loses the active power reserves of the conventional plants. And, second, it loses the instantaneous power reserves, because wind turbine generator sets are operated decoupled from the system frequency, which allows for aerodynamically efficient operation. In detail, the turbine's synchronous or asynchronous generators are connected to the grid via fast controlled power electronics [40–42].

Studies carried out by the Irish regulator have established that, with 60–70% of the total instantaneous power generated from WPPs, system frequency stability could be compromised [43].

In order to maintain system frequency stability in a network with an increasing share of wind power, wind turbines will have to take on more and more tasks of conventional power plants related to frequency control. This is reflected by the gradual development of more stringent requirements by system operators in regard to the integration of WPPs into network frequency control [44]. According to some system operators – for example, the Irish operator [45] – WPPs are already needed to provide power reserves. Also, future regulations will appear with the development of new requirements regarding synthetic inertia by WPPs [44].

Even though the power output of wind turbines depends on the unreliable wind speed, which is difficult to predict, and the generator set does not provide a passive instantaneous power reserve, there are methods by means of which WPPs can actually provide power reserves and thus participate in grid frequency control. Such abilities will be crucial for the successful integration of WPPs into the grid.

This chapter presents a review of selected European grid codes and future trends regarding the tasks of WPPs related to participation in frequency control. It also offers a literature review of the proposed methods for enabling wind turbines to provide active power reserves. Furthermore, the possibilities for wind turbines to provide instantaneous reserves are discussed.

3.1 A Review of European Grid Codes Regarding Participation in Frequency Control

Due to the island location of Ireland and the United Kingdom (UK), frequency control is a particularly challenging task in these electrical networks, since they do not have access to the large power reserves in the interconnected networks of continental Europe. Thus, the requirements for WPPs are significantly stricter in these networks than in the continental grid. However, the rising share of wind power will also lead to stricter requirements in continental Europe.

Accordingly, this section first gives the definition and nomenclature for different types of active power reserves. Second, deployment times of power reserves for selected European grid codes are depicted. Then, particular requirements WPPs regarding frequency control according to the grid codes of Ireland and the UK are presented. Finally, future trends based on the latest ENTSO-E Network Code [44] are discussed.

3.1.1 Nomenclature and the Definition of Power Reserves

Power reserves can be defined as the additional active power (positive or negative) that can be delivered by a generating unit in response to a power imbalance in the network between generation and consumption. Four different reserve levels can be defined: *instantaneous, primary, secondary*, and *tertiary* power reserves. This terminology is widely accepted; however, the nomenclature can vary from one country to another. The following paragraphs provide the definition of each power reserve.

Instantaneous power reserves refer to the physical stabilizing effect of all connected synchronous generators due to their inertia. In the event of a generation drop in the network, the instantaneous reserves balance the power due to this stabilizing and passive effect. Their electric power P_{elect} increases rapidly, which provokes an electromechanical imbalance in the generator set according to

$$P_{mech} - P_{elect} = J\omega_g \frac{d\omega_g}{dt}, \tag{3.1}$$

where P_{mech} is the mechanical power developed by the generator, J is the moment of inertia referred to the generator shaft, and ω_g the mechanical speed of the generator. The electrical rotational speed of the generator ω_r is deduced from the number of pairs p and ω_g as

$$\omega_r = \omega_g \frac{p}{2}. \tag{3.2}$$

As a result of the power imbalance, the rotational electrical speed ω_r decreases. This reduction also takes place in the interrelated frequency of the system. The rate of change of the system frequency (ROCOF) depends on the amount of available instantaneous power reserves and thus on the inertia of the system. Low levels of system inertia – that is, high levels of ROCOF – can provoke the tripping of sensible loads, generating units, and relays (implemented to avoid islanding), thus affecting the system frequency stability.

For studies related to power systems, it is a common practice to define the inertia constant H. The inertia constant, in seconds, determines the duration for which the generating unit may theoretically provide its rated power using only the kinetic energy stored in its rotating parts. It can be mathematically expressed as half of the mechanical acceleration time constant τ_{acc} (in seconds):

$$H = \frac{1}{2}\tau_{acc} = \frac{1}{2}J\frac{(\omega_g^{nom})^2}{P_{nom}^{total}}, \tag{3.3}$$

where ω_g^{nom} is the nominal mechanical generator speed in rad/s, P_{nom}^{total} is the nominal power of the generating unit, and J is the moment of inertia in kgm^2, referred to the generator shaft. For further details regarding the definition of the inertia constant, see Kayikci and Milanovic [39] and Ackermann [46]. It is important to note that wind turbines do not have inertia from the electrical system point of view, since the rotor is not synchronized with the network but connected through power electronics. According to Lalor, Mullane, and O'Malley [41], the ROCOF will increase under system frequency disturbances as wind generation displaces conventional generation, due to the decoupling of the rotor of the generators due to their fast controlled power electronics [39]. However, this does not happen with squirrel-cage asynchronous generators, since they provide a naturally inertial response, being directly connected to the grid [47].

The *primary reserve* is intended to be the additional capacity of the network that can be automatically and locally activated by the generator's governor after a few seconds at most of an imbalance between demand and supply of electricity in the network [48]. The aim of the primary reserve is to quickly balance the consumed and generated power in the system and thus stabilize the frequency at a certain level. These primary reserves are typically activated automatically by frequency droop controllers of generating units, building up the so-called *primary frequency control*. Primary reserves must be delivered until the power deviation is completely offset by *secondary* or *tertiary reserves*. With the aim of harmonizing the nomenclature related to power reserves, in June 2012 the European Network of Transmission System Operators for Electricity (ENTSO-E), which is the association of transmission system operators (TSOs) in continental Europe, defined primary reserves as *frequency containment reserves* [49]. Prior to the publication of this document, in the ENTSO-E's *Operational Handbook* [48], this reserve level had been set as the *primary control reserve*. This book considers both above-mentioned publications (due to their complementary and nonexclusive character), in order to depict the ENTSO-E's recommendations regarding system frequency control and reserve power levels.

Secondary reserves are activated to restore the rated frequency of the system, to release primary reserves, and to restore active power interchanges between control areas to their setpoints [48]. They are activated by the TSOs by modifying the corresponding active power setpoints of the generating units within each respective control area. In June 2012, ENTSO-E proposed the term *frequency restoration reserves* to define secondary reserves [49].

Finally, the aim of *tertiary reserves* (or *replacement reserves*, according to the most recent the ENTSO-E proposal [49]) is to replace the secondary reserves and restore the frequency to its rated value if the secondary reserves are insufficient. Furthermore, the tertiary reserves are used for economic power dispatch [50], considering system constraints such as the current limits of the transmission lines. These reserves are activated manually and centrally at the TSO's control centers in the event of observed sustained activation of secondary reserves or in anticipation of a response to expected imbalances [48].

The general rules and technical recommendations regarding reserve power levels and their associated control performance are set out in the ENTSO-E's *Operational Handbook* [48]. Despite these general guidelines, specific rules for the provision and control of power reserves for grid frequency control must be determined by the TSOs in their own grid codes.

3.1.2 *The Deployment Sequence of Power Reserves for Frequency Control*

Figure 3.1 depicts the principal behavior of the system frequency after a sudden lack of power generation in the network, with all the above-mentioned power reserve levels being involved. Some annotations on their activation are included, according to the ENTSO-E [48].

In the event of a generator trip in the network, the system frequency starts to drop and thus the instantaneous reserves of the synchronous generating units help to restore the power balance in the network due to the stabilizing effect of the inertia. Then, as soon as a certain frequency level is reached, the primary power reserves are activated; the driving mechanical power of the generators increases. As equilibrium between the developed electrical and mechanical power in the generators is achieved, the system frequency stops dropping; in other words, the frequency nadir is achieved. Next, a further increase of the primary reserve power accelerates the generator sets and the system frequency increases up to a new steady state below its rated value.

The slope of the primary power – frequency droop characteristic has a major influence on the achieved frequency nadir and on the frequency stabilization level [39,51]. In this sense, it can be concluded that the slower the response of the governor of the generating unit, the lower is the frequency nadir.

The activation of the secondary reserves allows the normal operating frequency levels to recuperate and thus deactivates the primary reserves. The secondary reserves are operated until they are fully replaced by the tertiary reserves. Proper explanations of the parameters presented in Figure 3.1 are as follows [48]:

- f_{min} **and** f_{max}. The minimum and maximum expected instantaneous frequencies after a reference incident (loss of generation or loss of load), assuming predefined system conditions.

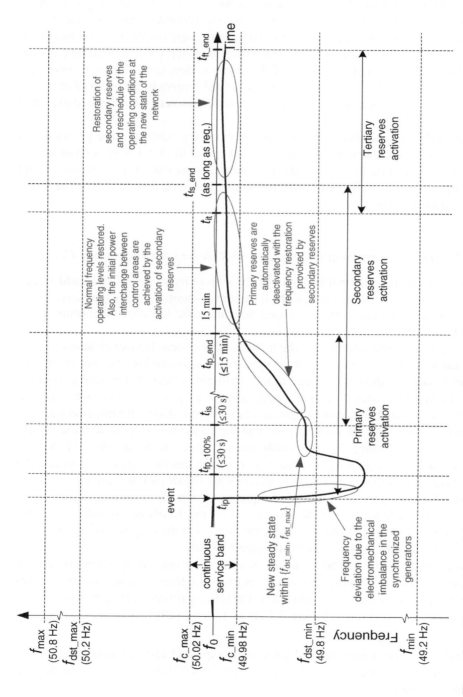

Figure 3.1 The definition of the concepts: particular values of time frames and frequencies follow the ENTSO-E recommendations. *Source:* Adapted from ENTSO-E, 2009 [48].

- f_{dst_min} **and** f_{dst_max}. The minimum and maximum steady state frequencies. They define the tolerance band for the quasi-steady state system frequency after the occurrence of a reference incident, assuming predefined system conditions. Outside this interval, all the available primary reserves remain activated. This means that the droop controllers of all of the units that are providing primary reserves should be set to deploy all of their contracted or obligatory primary reserve capacity.

- f_{c_min} **and** f_{c_max}. The limits of the frequency deadband for the activation of primary reserves. These limits define an interval in which primary reserves do not need to be activated. This allowed frequency deviation usually corresponds to the accuracy of the frequency measurement and the insensitivity of the controller. That notwithstanding, a greater deadband is also permitted in accordance with the TSO.

- t_{ip}, t_{is} **and** t_{it}. The maximum starting time for the activation of the primary, secondary, and tertiary reserves from the event detection time.

- $t_{fp\_50\%}$ **and** $t_{fp\_100\%}$. The maximum deployment times for 50% and 100%, respectively, of the total primary reserves from the event detection time.

- $t_{fs\_100\%}$ **and** $t_{ft\_100\%}$. The maximum deployment times for 100% of the total secondary and tertiary reserves, respectively, from the event detection time.

- t_{fp_end}, t_{fs_end} **and** t_{ft_end}. The minimum capabilities of actuation of the primary, secondary, and tertiary reserves, respectively.

Accordingly, Table 3.1 presents specific values of the above-listed frequency levels and time frames from the ENTSO-E's recommendations in its *Operational Handbook* [48], as well as particular data from selected European grid codes. It must be noted that the values indicated apply to conventional generating units participating in system frequency control.

As can be noted in Table 3.1, the specified frequency levels in both the German [52] and the Spanish [53, 54] Grid Codes mainly match the ENTSO-E's recommendations [48]. In fact, the deployment sequence for the power reserves does not differ essentially from one country to another within the interconnected continental European networks. This permits us, for instance, to conclude that in continental European networks, the primary reserves need to be fully activated 30 s at most after the detection of a frequency deviation.

The deployment sequence for power reserves in islanded grids is quite different from those for continental European networks. For instance, the Irish Grid Code requires a faster response time in full activation of power reserves (see Table 3.1). Despite the fact that we were unable to find a particular value for the response time in full activation of primary reserves in the Irish Grid Code, the activation time for secondary reserves is set to at most 5 s after the detection of a frequency deviation, which is a shorter time than that provided for the full activation of primary reserves in continental European networks. Furthermore, it is remarkable that the concept of secondary reserves does not exist in the UK's electrical network, which hence distinguishes itself from the

Table 3.1 Parameters extracted from the ENTSO-E's recommendations and some European grid codes for conventional generating units participating in system frequency control.

Parameter	ENTSO-E [48] (2009)	German Grid Code [52] (2007)	Spanish Grid Code [53, 54] (1998, 2009)	Irish Grid Code [45] (2011)	UK Grid Code [55] (2012)
f_0	50 Hz	50 Hz	50 Hz	50 Hz	50 Hz
$\{f_{c\_min}; f_{c\_max}\}$	{49.98;50.02} Hz	{49.98;50.02} Hz	{49.98;50.02} Hz	{49.985;50.015} Hz	{49.985;50.015} Hz
$\{f_{dst\_min}; f_{dst\_max}\}$	{49.8;50.2} Hz	{49.8;50.2} Hz	{49.8;50.2} Hz	{49.5;50.5} Hz	{49.5;50.5} Hz
$\{f_{min}; f_{max}\}$	{49.2;50.8} Hz	{49.2;50.8} Hz	{49.2;50.8} Hz	{49.0;51.0} Hz	{49.2;50.8} Hz
t_{fp}	A few seconds after detecting a frequency deviation of ±20 mHz[b]	A few seconds after detecting a frequency deviation of ±20 mHz	A few seconds after detecting a frequency deviation of ±20 mHz	0 s (or with a frequency deviation of ±15 mHz)	0 s (or with a frequency deviation of ±15 mHz)[c]
$t_{fp\_50\%}$	≤15 s	–	≤15 s[d]	–	–
$t_{fp\_100\%}$	≤30 s	≤30 s	≤30 s[e]	–	≤30 s[f]
t_{fp_end}	≥15 min	≥15 min	As long as required	≥30 s	≥30 min[g]
t_{is}	≤30 s	≤30 s	–	≤5 s	–
$t_{is\_100\%}$	≤15 min	≤15 min	300 s ≤ 500 s[h]	≤15 s	–
t_{is_end}	As long as required	As long as required	≥15 min	10 min	–
t_{tt}	In TSO's decision	In TSO's decision	In TSO's decision	In TSO's decision	In TSO's decision
$t_{ft\_100\%}$	A short time	A short time	≤15 min	–	–
t_{ft_end}	–	–	≥2 h 15 min	–	–

[a] According to the corresponding document: section A-S2.3, "Physical deployment times."

[b] According to the German Grid Code, section 5.2.2, page 39 [52], and referring to the frequency control, "the TSO shall use primary control power in accordance with the rules of the UCTE-OH Policy 1," which is the document referred to in the first column of the table [48].

[c] According to the UK Grid Code, section CC.A.3.4 [55], and referring to the activation of primary reserves, "the active power output should be released increasingly with time over the period 0 to 10 seconds from the time of the start of the frequency fall."

[d] The deployment time from 50 to 100% of total primary reserve rises linearly from $t_{fp\_50\%}$ to $t_{fp\_100\%}$.

[e] If the network frequency variation is less than 100 mHz, $t_{fp\_100\%}$ =15 s.

[f] This value corresponds to the maximum deployment time of the so-called primary response capability of a generator unit operating in frequency-sensitive mode. For more details, refer to Section 2.3.

[g] This value corresponds to the minimum capability of actuation of the so-called secondary responses of a generating unit operating in frequency-sensitive mode. For more details, refer to Section 2.3.

[h] Response time constant of 100 s of a first-order type system.

rest of the European networks considered. This is because frequency control in this network is done just by using the primary governors of the generators.

Apart from setting the deployment times for the power reserves, the regulations also specify the power reserve needs for each control area of the network, so that the stability of the system can be ensured. The required primary reserves in the synchronized European network to stabilize system frequency are defined on the basis of the so-called referent incident. This referent incident is the maximum expected power deviation between generation and consumption in a network. The primary reserves must be able to offset the power imbalance caused by a reference incident. For continental Europe, this referent incident, or primary reserve need, is estimated at 3000 MW [48]. This total power reserve is allocated throughout the network, attending to the specificities of the grid of each country.

Usually, the providers of power reserves (mostly primary and secondary power reserves) are power plants with relatively high ramp power rates and short time responses, due to the required fast dynamics for regulating their power output for tasks related to frequency control. In this regard, open- and combined-cycle gas turbine–based power plants, as well as hydropower plants, are the most suitable technologies.

In fact, hydropower is a mature technology, which nowadays represents more than 80% of the total renewable energy generated worldwide. In Germany, for example, the total hydropower capacity installed is very high, at around 4.3 GW, providing principal power reserves to ensure the stability of the continental European network. Some figures illustrating the high ramp power rates of such installations have been published by First Hydro Company [56]. This company reports the main features of a pumped hydroelectric storage installation able to move from 0 to 1320 MW power output in just 12 s, managing six motor-generators of 330 MW activated by reversible Francis water turbines. The major drawback of hydropower is the environmental aspects resulting from the required civil construction work, which affects the natural flow of rivers. The scarcity of suitable sites for installation of the power plant, which are commonly natural and in areas that are difficult to access, and the need to install new transmission lines to transport the electricity generated to the consumers, are also viewed as important environmental impacts to be taken into consideration.

Gas turbine–based power plants are also very flexible, so they can be activated quickly to provide power reserves when needed. They are also used for other purposes, such as providing the service of peak shaving during periods of high demand. The fact that these power plants are non–CO_2 neutral is a major drawback for this technology in comparison with hydropower plants. On the other hand, both the capital expenditure and the operating and maintenance expenditures of such installations are lower than in the case of hydropower-based plants, as depicted in [57].

Finally, it is important to note that the valorization schemes for the provision of power reserves can also vary from one country to another, according to the various grid codes and rules for the differing electrical markets. For instance, according to the German Grid Code [52], it is mandatory for just conventional generating units with a rated power greater than 100 MW to provide primary reserves to the system.

Smaller generators may also be employed upon agreement with the TSO [52]. The participation in this service is remunerated, as is the provision of secondary and tertiary reserves. Unlike the German Grid Code, the Spanish Grid Code [53,54] specifies that all synchronized generators must provide primary reserves, but this participation is not paid for. Only the provision of secondary and tertiary reserves is market-regulated.

Each country sets different rules for the operation of its electrical energy markets. Apart from the day-ahead and intraday markets, in which the majority of the electricity consumed daily is negotiated, there are other markets for the provision of power reserves, as well as for the provision of other ancillary services to the network. In this regard, and following the example of Spain, the secondary and tertiary power reserve needs are negotiated in the so-called secondary and tertiary markets. The providers of these reserves participate in the corresponding markets by presenting their bids, specifying the offered reserve level (in MW) with the corresponding price (in € /MW), for each operating period of the following day. Thus, the assignation of the providers of the reserves is done in advance, so as to ensure the provision of sufficient reserves for proper operation of the network during the day.

European trends in this regard are devoted to the integration of markets such as the day-ahead, intraday, and others for the provision of ancillary services, as this is intended to be a source of flexibility for the future European electrical networks [58]. Flexibility is essential to enhance the accommodation of growing amounts of renewable (and thus variable) generation. Other measures that will transform the markets of the future are, for instance, the reduction of the common hourly basis operation of the electrical markets down to quarter-hourly or even a real-time basis. This will reduce forecasting errors for the generation and demand in the network. Forecasting errors compromise the proper dispatch of the generators and the determination of the required power reserve levels.

Previous paragraphs refer to the provision of power reserves by conventional generating units in different European countries. This serves as a starting point for the introducion of the specificities for the provision of power reserves by WPPs in the following sections.

The grid codes of islanded European networks, such as those for the networks of Ireland and the UK, set specic requirements for renewable generating units regarding the provision of primary reserves in case of both underfrequency and overfrequency events. In contrast, in continental European networks with a high penetration of renewable-based power plants, as in the case of Germany and Spain, the current grid codes do not require renewable generating units to provide primary reserves for their participation in tasks related to frequency control. Nevertheless, renewable generating units are needed within the German network to reduce their output for frequency control purposes at system frequencies above 50.2 Hz[1] and they can also provide primary reserves, but are not required to do so [59].

---

[1] In the transmission code, section 3.3.13.3, page 35 in the German version, figure 3.4.

Accordingly, the subsequent sections set out the most relevant aspects regarding system frequency control support by WPPs in keeping with the Irish and UK Grid Codes [45,55]. A brief note about secondary reserves according to the Irish regulations are also included, as this is considered particularly interesting for the purpose of this chapter.

3.1.3 A Detailed View on the Requirements for WPPs in the Irish Grid Code

The Irish regulations detail, amongst other contents, a specific set of requirements for WPPs regarding some aspects concerning their controllability and behavior during grid disturbances and their participation in frequency and voltage control. There are some differences between the ENTSO-E [49] and the Irish terminology regarding power reserves. The Irish Grid Code adopts the term *operating margin*. This represents the power reserve to be sustained to meet the expected system demand for limiting and correcting deviations from the system frequency. The *operating margin* includes the so-called *operating reserve*, the *replacement reserve*, the *substitute reserve*, and the *contingency reserve*. The operating reserve is defined as "the additional power output provided by generating units realizable in real time operation to limit and correct system frequency deviations to an acceptable level." This operating reserve consists, in turn, of the *primary operating reserve*, the *secondary operating reserve*, the *tertiary operating reserve band 1*, and the *tertiary operating reserve band 2*. Each of these operating reserves applies over different time frames up to 20 min following an event.

Moreover, and continuing with the description of the terminology differences, frequency control is carried out by means of using the operating reserves and occurs on two timescales: *primary frequency control* and *secondary frequency control*.

Figure 3.2 summarizes the equivalences between the ENTSO-E and Irish terminologies regarding power reserves. Moreover, it also depicts the equivalences with the different terminologies adopted by the rest of the European grid codes considered in this chapter. According to the Irish terminology, WPPs are required to participate in system frequency control with the provision of primary reserves and secondary reserves.

As set out in Table 3.1, primary reserves take place in the period of up to 30 s after the detection of a frequency deviation. They are achieved by automatic corrective responses, which include governor droop actions of generators and automatic load shedding. Only power plants based on synchronous machines, with a rated power greater than 60 MW, have to regulate their primary reserves in response to a frequency deviation between 49.8 Hz and 50.2 Hz. In contrast, a wind farm control system has to present the capabilities shown in Figure 3.3 regarding the activation of primary reserves. Points P_A to P_E, and f_A to f_E, are determined by the TSO before the start of operation of the unit. The activation of the power reserves is carried out by automatic

ENTSO-E terminology		Spain	Germany	UK	Ireland
2009	2012	1998, 2009	2007	2012	2011
Primary reserves: Reserves that are automatically and locally activated while a frequency deviation occurs	**Frequency containment reserves**	**Primary reserves**	**Primary control power**	**Reserves for primary, secondary, and high-frequency response**	**Primary operating reserves** Time frame: 30 s from event
Secondary reserves: Reserves that are automatically or manually activated by TSO a certain time after a frequency deviation occurs, and required for generating units of a control area	**Frequency restoration reserves**	**Secondary reserves**	**Secondary control power**	*This concept does not exist in the UK Grid Code*	**Secondary op. Reserves:** fully available from 15 s to 90 s after event **Tertiary op. Reserves:** - *Band 1:* fully available from 90 s to 5 min after event **Tertiary op. Reserves:** - *Band 2:* fully available from 5 to 20 min after event
Tertiary reserves: Reserves that are manually activated by TSO in case of observed sustained activation of secondary reserves or in response to an expected imbalance	**Replacement reserves**	**Tertiary reserves**	**Minutes reserve power**	**Operating and contingency reserves**	**Replacement reserves:** fully available from 20 min to 4 h after event **Substitute Reserves:** fully available from 4 h to 24 h after event **Contingency reserves:** fully available from 24 h to "a limited time scale" after event

Reserves taking part in "frequency control" under the timescales...

"primary frequency control" (up to 30 s after event)

and "secondary frequency control" (from 5 s to 10 min after event)

Requirements for wind power plants and conventional synchronized generating units with rated power greater than 2 MW

Figure 3.2 The equivalences between the ENTSO-E (2009) [48], ENTSO-E (2012) [49], and Irish [45] regulations. Equivalences with the Spanish, German, and UK Grid Codes are also depicted.

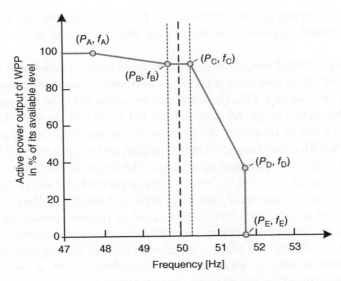

Figure 3.3 The droop characteristic for the activation of primary reserves according to the requirements set out by the Irish regulations for WPPs. *Source:* Adapted from EirGrid, 2013 [45].

local controllers. As can be noted, at the rated system frequency, a WPP is required to feed in less than its available active power. This derated operation allows the WPP to provide both positive and negative power reserves; that is, to ramp power both up and down in response to deviations in the system frequency. The primary reserves should be activated immediately after the detection of a frequency deviation from the specified deadband, without any control signal from the TSO.

Secondary reserves come into play in the time range from 5 s up to 10 min after the frequency deviation is detected. Secondary frequency control is carried out by a combination of automatic and manual actions (dispatch instructions from the TSO). The active power setpoint comes from the TSO, and a time delay of up to 10 s from receiving this setpoint is allowed for their activation.

As for active power ramp rates for the activation of both primary and secondary reserves, the response rate of each available wind turbine must be at least 1% of its rated capacity per second. Moreover, the TSO limits the active power ramp rate to the WPP. In this sense, it has to be possible to vary the active power ramp rate between 1 MW/min and 30 MW/min.

3.1.4 A Detailed View on the Requirements for WPPs in the UK Grid Code

Each WPP (both onshore and offshore) with a registered capacity over 50 MW must be capable of participating in frequency control by continuously adjusting its active power output. This active power control can be performed by applying two operating modes, the so-called *frequency-sensitive mode* and the *limited frequency-sensitive mode*.

In the latter operational mode, the generating units must be capable of maintaining a constant level of active power output for system frequency changes between 49.5 Hz and 50.5 Hz. In the case of WPPs, the active power output has to be independent of the system frequency in this range. Moreover, from below 49.5 Hz to 47.0 Hz, a possible active power drop due to frequency decay must not exceed 5%. This operating mode applies to WPPs with a rated capacity both less than and greater than 50 MW.

Participation in frequency control according to the frequency-sensitive mode is part of the formulated ancillary services, which are categorized into the so-called *system ancillary services* and *commercial ancillary services*. System ancillary services refer to the provision of mandatory services in respect of reactive power and frequency control support. Commercial ancillary services include aspects related, for example, to the fast start capability, the black start capability, and the programmed tripping of generating units to prevent abnormal system conditions such as overvoltage and system instability caused by, for instance, system faults.

Therefore, by performing in the frequency-sensitive mode, WPPs provide a system ancillary service. Only WPPs with a rated power greater than 50 MW must have the capability to provide this ancillary service, and thus no longer be operated in limited frequency-sensitive mode, but in frequency-sensitive mode upon instruction from the TSO.

The term "frequency-sensitive mode" is the generic description of a mode of operation that includes the provision of the *primary response*, and/or the *secondary response* and/or the *high-frequency response*. The so-called primary and secondary responses refer to negative frequency deviations, while the high-frequency response refers to positive frequency deviations. These frequency response capabilities are activated by automatic controllers in the generating unit. Consequently, and according to the terminology used in this chapter, the primary, secondary, and high-frequency response capabilities can be intended as primary reserves. Each response capability must be tested by inducing a ramp to the frequency control device from 0 to a 0.5 Hz change over a period of 10 s. This frequency deviation must be sustained thereafter.

The primary response capability of a generating unit is the minimum increase in the active power between 10 and 30 s after the start of the induced frequency deviation ramp to the controller. The secondary response capability is the minimum increase in the active power output between 30 s and 30 min after the activation of this ramp. Finally, the high-frequency response capability is the decrease in active power output within 10 s after the induction of, in this case, a frequency ramp with a positive slope. These concepts are depicted in Figure 3.4.

Being operated in frequency-sensitive mode, WPPs should not extract the maximum available power from the wind but instead have to be derated, so that they can ramp up their output and down according to the frequency of the network. The minimum

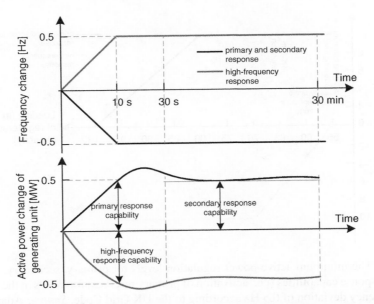

Figure 3.4 A representation of the primary, secondary, and high-frequency response capabilities according to UK Grid Code. *Source:* Adapted from National Grid plc, 2012 [55].

power output change required of a WPP operated in frequency-sensitive mode can be found in Figure 3.5.

As shown, the change in the generation level depends on the actual loading of the unit. A generating unit must be capable of providing a frequency response at least up to the indicated boundaries, depending on its loading. Since the indicated power change levels correspond to a frequency deviation of 0.5 Hz, a directly proportional change in the power level has to be determined for smaller frequency deviations; that is, according to a power – frequency droop control. For instance, if we assume that the wind speed is such that the wind turbine could generate 75% of its registered capacity (i.e., the loading is 75%), it should be operationally derated enough to be able to ramp up its power output by 10% of its registered capacity for primary and secondary responses (see Figure 3.5).

Moreover, in Figure 3.5, two operational limits are highlighted: the *designed minimum operating level* and the *minimum generation level*. The latter defines the minimum stationary part-load level at which it must be possible to sustain the generating unit. This level should not exceed 65% of the rated power (for further explanation, see [55], page CC-68). For instance, a wind turbine, or a thermal power plant, must be capable of working at 65 or 60% of its rated capacity in steady state. The former concept, the designed minimum operating level, bounds the minimum generation level at which the generating unit must provide a high-frequency response; that is, at which it must activate negative primary reserves at grid frequencies above 50.0 Hz.

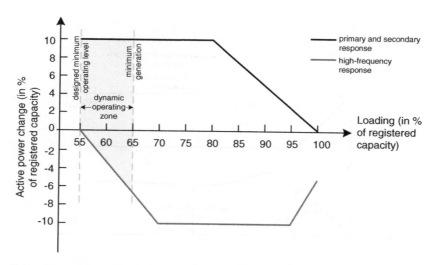

Figure 3.5   The minimum active power regulation levels for primary, secondary, and high-frequency response capabilities (i.e., activation of primary reserves) for WPPs in the event of a system frequency deviation of 0.5 Hz according to the UK Grid Code. *Source:* Adapted from National Grid plc, 2012 [55].

The generating unit would also have to provide a high-frequency response under the designed minimum operating level, but only considering frequencies above 50.5 Hz. Besides, it should be noted that the deadband of the frequency control devices in frequency-sensitive mode must be ±0.015 Hz at most.

From Figure 3.5, and as previously noted, we can conclude that in frequency-sensitive mode, similar to the Irish case, a continuous power derating should be applied to WPPs in order to be able to transiently increase their output according to the requirements for the provision of primary reserves (provided that no energy storage devices are included in the WPP).

3.1.5   Future Trends Regarding the Provision of Primary Reserves and Synthetic Inertia by WPPs

In order to harmonize the connection requirements for the generating units of the European power system, the ENTSO-E presented its first network code [44] in June 2012 (see Section 3.1.1). The code applies to all grid-connected generators. Generating units are classified into different types according to the voltage level at their grid connection point and their rated power and region. Despite the fact that the code does not refer explicitly to the terms "primary reserves," "secondary reserves," and "tertiary reserves," it details specific requirements for generating units regarding their participation in frequency control.

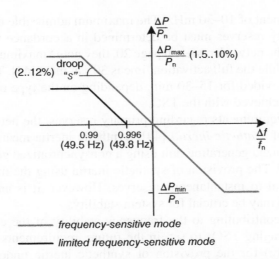

Figure 3.6 The active power – frequency response droop characteristic according to the ENTSO-E network code. *Source:* Adapted from ENTSO-E, 2013 [44].

The most restrictive requirements regarding frequency control are borne by the generating units categorized as type C. This category covers, for instance, WPPs with a rated power above 50 MW in continental Europe, above 10 MW in the UK, and above 5 MW in Ireland. These WPPs must be equipped with a power control system for frequency response. There are two operational modes for this control system: the *limited frequency-sensitive mode* and the *frequency-sensitive mode.*[2]

In the latter operating mode, the control system is in charge of ramping the active power output of the generating unit up and down in the event of over- and underfrequency, according to a power – frequency droop characteristic. The *limited frequency-sensitive mode*, as a particular case, just requires the generating unit to increase its power output while the frequency lies between 49.5 and 49.8 Hz. These concepts are depicted in Figure 3.6.

As shown, in *frequency-sensitive mode* a WPP must be capable of regulating its power output according to the system frequency within a given range around the currently available power. The required range can be defined between 1.5 and 10% of the nominal power of the plant. The droop characteristic is saturated at predefined frequency levels, which must be determined in accordance with the relevant TSO. The droop characteristic must present a slope of between 2 and 12% and could include a deadband up to 0.5 Hz (in accordance with the relevant TSO). Finally, taking into account inevitable frequency measurement errors, there is a tolerance for the

---

[2] The ENTSO-E definition of these terms is differs from that in the UK regulations. The ENTSO-E code requires a WPP to ramp up its active power output during underfrequency events, but the UK regulations (see Section 3.1.4) require the WPP to maintain a constant active power output.

frequency measurement of 10–30 mHz. The maximum admissible time delay for the activation of primary reserves must be determined in accordance with the relevant TSO (in Table 4 of the network code, on page 20, they note "maximum 2 s" for type C generating units), while the full activation time is 30 s at most [44]. This active power surplus has to be provided for 15–30 min, depending on the type of generating unit and the agreement achieved with the TSO.

Apart from the requirements regarding primary reserves, the network code introduces the concept of *synthetic inertia* [44]. Synthetic inertia means replicating the inertia of a synchronized generating unit using a nonsynchronized generating unit; in other words, a WPP. The provision of synthetic inertia using the methods available today is not identical to instantaneous reserves. However, it is an approximation, which is helpful and may be crucial for system stability.

With the aim of contributing to the frequency stability of the electrical system, ENTSO-E is encouraging TSOs to set, in the future, requirements for WPPs (both onshore and offshore) for the provision of synthetic inertia under low-frequency events. In particular, this implies injecting power proportional to the "severity"[3] of the disturbance in a very short time (200 ms) [44].

Hydro-Québec TransÉnergie have recently carried out research on synthetic inertia needs [60]. The analysis quantifies the impact on frequency performance and stability of the inclusion of a new 2000 MW of wind power capacity in the Hydro-Québec transmission system. Simulation results show that in order to maintain the system frequency within its operating limits under a low-frequency disturbance (58.5 Hz) and thus to avoid automatic load shedding, wind farms have to participate in frequency regulation by providing synthetic inertia. It is worth highlighting that the minimum duration of the active power contribution for synthetic inertia is set at 10 s (taking a 6% active power increase into consideration) in order to expand the contribution beyond the frequency nadir. This is important, since ending the provision of synthetic inertia in the first few seconds of the frequency disturbance, before the frequency reaches a new steady state after the disturbance, could increase the frequency nadir further, as it can be viewed as a further loss of generation during the transient. The results of this research are consistent with those of Gonzalez-Longatt [61], which determine that the synthetic inertia provided by wind turbines can help to avoid the activation of underfrequency automatic load shedding, as higher-frequency nadirs are registered during a frequency disturbance.

In addition to synthetic inertia, another option for wind turbines to at least partly compensate for missing instantaneous reserves in the system is a very fast frequency response (according to a power – frequency droop characteristic) with a very short time response and time delay [62]. The next section will review and discuss methods for the provision of both primary frequency control and synthetic inertia.

---

[3] This requirement is not defined in further detail in the grid code [44].

3.2 Participation Methods for WPPs with Regard to Primary Frequency Control and Synthetic Inertia

3.2.1 Deloading Methods of Wind Turbines for Primary Frequency Control

Conventionally, wind turbines are operated at maximum aerodynamic efficiency, so that they can maximize the power extracted from the wind. In the partial-load region, the speed of the turbine is controlled by the regulation of the aerodynamic torque (or power) [63], leading to the so-called optimal torque rotor-speed curve. In particular, the optimal aerodynamic power is computed as follows:

$$P^* = K_{cp}(\beta) \times \omega_t^2,\tag{3.4}$$

where ω_t is the speed of the turbine, and K_{cp} is the so-called optimal aerodynamic torque coefficient, which depends on the aerodynamics of the turbine and the pitch angle β. Usually, this curve is implemented using a lookup table in the controller of the machine-side power converter of the wind turbine, leading the so-called maximum power tracking algorithm (MPT).

For maximum power generation, β is maintained constant at zero degrees in partial-load operation of the turbine. In contrast, in full-load operation, the generator torque is kept constant and the pitch control is activated in order to limit the power captured from the wind, and thus to ensure that it does not exceed the turbine rated power [64–67]. The pitch angle can be regulated by, for example, by a PI controller, which is governed by the difference between the turbine rated power and the actual measurement [68]. Controlling the rotor speed at constant torque can also be a possibility for driving this PI controller [69]. Furthermore, it is interesting to note that the pitch angle may also be varied cyclically in order to mitigate the mechanical loads of the turbine [70].

As noted in Section 3.1, WPPs are required to participate in primary frequency control by ramping their output up and down according to a power – frequency droop characteristic and over a certain period of time. This means that wind turbines either have to be operated in a deloaded mode or a suitable energy storage device has to be available. This chapter focuses on solutions for the first method. Various approaches can be found in the literature on the subject of deloading methods for wind turbines and they can be classified into two main categories: pitching techniques and overspeeding techniques [40, 71].

Both methods are based on the idea of achieving a suboptimal working point on the torque–rotor-speed curve of the turbine (or, analogously, on the power–rotor-speed curve). As can be noted in Figure 3.7, deloading is then realized by using C (overspeeding the turbine) or B (pitching the blades) as standard operating points and switching between point A and additional deloaded points.

While applying pitching techniques, the rotational speed of the turbine is kept constant (point B). Conversely, while applying overspeeding techniques, the turbine

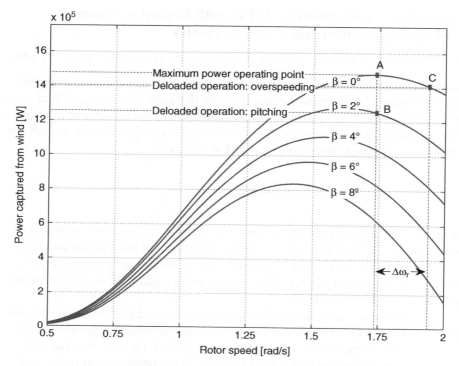

Figure 3.7 Power–rotor-speed curves for various values of the pitch angle and deloaded options for a 1.5 MW wind turbine (wind speed 10 m/s).

is operated at a higher rotor speed, while keeping the pitch angle constant (point C). One main difference between these two strategies is the fact that overspeeding can only be applied in variable-speed wind turbines (i.e., DFIG-based and full-converter-based wind turbines).

The application of underspeeding techniques is not recommended. A deceleration of the rotor from point A in order to reduce the active power output would lead transiently to an increase of the active power. This is due to the fact that the rotor releases kinetic energy. At the same time, when moving to point A from an underspeed level, the rotor would transiently consume active power for acceleration. Small-signal stability could be also endangered (unlike the situation with respect to pitching techniques) [72].

Depending on the wind speed level, either the one method or the other (pitching or overspeeding) is advantageous. Three different operating regions for wind turbines can be considered in this sense: the low, medium, and high wind speed ranges. In the low wind speed range, wind turbines operate at partial load. Thus the rotating speed of the turbine does not reach its rated value at any time, allowing the application of overspeeding techniques. In the medium wind speed range, wind turbines mostly operate at partial load but they can achieve their rated rotational speed transiently,

Table 3.2 Considerations regarding deloading strategies of wind turbines.

Overspeeding techniques:
- Method preferably applied below rated wind speed levels [71].
- Wind speed measurements are usually required for determining the maximum power that a wind turbine could extract from the wind while applying the deloaded power – rotor speed curves. The accuracy and reliability of this wind speed measurement are crucial [73].
- Considering the DFIG, the percentage of power transmitted through the set of back-to-back power converters becomes greater with greater values of slip. This limits the overspeeding of the wind turbine in order not to overcome the ratings of the power converters [71, 74].
- Danger of excessive mechanical stress in the rotor shaft due to the fast deloading through the fast torque control for speed regulation. Need to apply limits to the rate of change of torque [72].
- Rotor speed regulation is only possible in variable-speed wind turbines.
- Due to the inertia of the rotor, the power ramping is not linear.

Pitching techniques:
- Method preferably applied above the rated wind speed [71].
- Excessive pitch control actions may lead to wear and tear of the mechanism [67, 72]. Moreover, pitch angle regulation could affect the fatigue life of the blades, as it affects their dynamic loads [69, 70].
- Larger time responses than in overspeeding techniques due to pitch servo time delays [74].
- Usually, no wind speed measurements are required [75].

and thus the pitch controller may be activated. In this region, a combination of overspeeding and pitching techniques may be a good option for deloading the wind turbine. And finally, the in high wind speed range, pitch control becomes a key factor for both limiting the power extracted from the wind in order not to exceed the ratings of the generator and for applying deloading strategies. Instead of only limiting the power, pitch control can be used for ramping power up and down. Overspeeding strategies generally do not fit well in this region. Apart from the limitations regarding wind speed for applying the above-mentioned deloading strategies, further related considerations can be found in the literature that must be taken into account for application of the deloading strategies mentioned. A brief summary of some of these is presented in Table 3.2.

The following subsections deal with a literature review regarding the application of each of the above-mentioned strategies.

3.2.1.1 The Control Basis of Overspeeding and Pitching Techniques for the Deloaded Operation of Wind Turbines: An Example

Figure 3.8 visualizes a general example for a controller structure that allows deloading and participation in frequency control by means of pitching and overspeeding techniques. It serves as the basis for the literature overview of the proposed methods. Considering low wind speeds and rotor speeds below rated, deloaded operation is

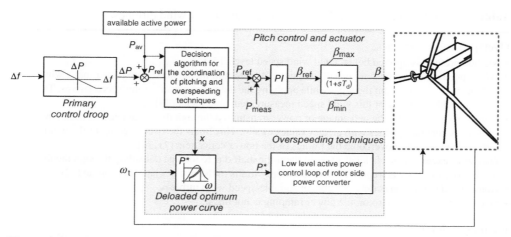

Figure 3.8 An example of a control scheme for a wind turbine for primary frequency control support. It includes the primary frequency control droop, the pitch control, and the rotor speed control.

preferably carried out by means of overspeeding techniques (i.e., by affecting the power reference to the power converters of the wind turbine). For wind speeds near or above nominal, when the maximum rotor speed is reached, the pitch controller is additionally activated.

As shown in Figure 3.8, the deloaded power reference P_{ref} of the wind turbine can be obtained from the sum of the available active power of the turbine (which depends on the wind), P_{av}, and the output of a primary frequency control droop ΔP. The computation of P_{av} can be carried out using the internal signals of the turbine and will not be discussed in detail here.

A decision algorithm coordinates the activation of the pitching and overspeeding techniques. It transfers the reference P_{ref} to the pitch control when it is being used, and it also translates P_{ref} and P_{av} into the requested power margin reference x, which is the reference value for the applied overspeeding technique. It is important to note that even while regulating the pitch angle, it is still necessary to control the power extracted through the converters of the wind turbine, so it is still necessary to determine a reference x consistent with the reference P_{ref} for the pitch control.

When deloading the wind turbine through pitching techniques, the pitch angle is driven by a PI controller, which tries to minimize the difference between the measurement of the generation level and the deloaded power reference P_{ref}.

The so-called deloaded optimum power curves are the key aspect for the implementation of overspeeding techniques. The following paragraphs explain the basis of this element; see Figure 3.9. Applying the conventional control approach according to Equation (3.4), the target operating points lie on the optimum power curve in the power–omega (P–ω) diagram. In contrast, for deloading and participation in frequency

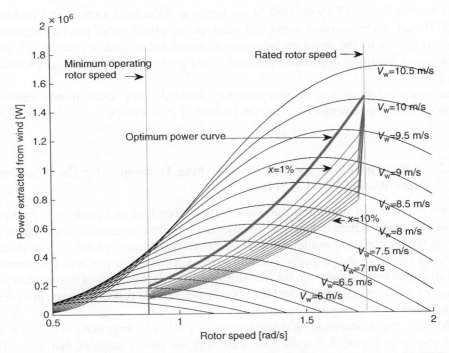

Figure 3.9 Deloaded optimum power curves for the deloaded operation of a 1.5 MW DFIG-based wind turbine.

control, the target operating points lie on a deloaded curve, which corresponds to the requested power margin x. The power curves can be precalculated. Thus, the power reference to the generator controller is retrieved from the measured turbine speed ω_t, the required power margin, and the power curve diagram.

The principle of deloaded optimum power curves is adopted in de Almeida, Castronuovo, and Peças-Lopes [73], de Almeida and Peças-Lopes [76], Zertek, Verbic, and Pantos [77, 78], Ramtharan, Ekanayake, and Jenkins [79], and Zhang *et al.* [80]. The deloaded power curves as depicted in Figure 3.9 can be precalculated as follows. The active power setpoint of the wind turbine $P_{del_{opt}}$ is related to the maximum available power P_{opt}, which is computed from Equation (3.4), and the required power margin x, as follows:

$$P_{del_{opt}} = P_{opt}(1 - x). \tag{3.5}$$

For this desired power level, a corresponding suboptimal power coefficient $C_{P_{del}}$ can be computed for a specific wind speed as follows:

$$C_{P_{del}} = \frac{P_{opt}(1 - x)}{0.5\rho A v_w^3} = C_{P_{del}}(v_w, \omega_t, \beta). \tag{3.6}$$

The power coefficient of a wind turbine is a function of the pitch angle, wind, and rotor speed. Thus, for a given pitch angle and wind speed, a rotor speed can be determined that corresponds to the required suboptimal power coefficient. By means of these operations, a family of deloaded optimum P–ω curves can be defined, as plotted in Figure 3.9.

The torque reference can be also computed instead of the active power reference by applying torque–omega (T–ω) curves instead of P–ω curves.

3.2.1.2 A Review of Overspeeding and Pitching Techniques for the Deloaded Operation of Wind Turbines

The previous section presented the basis for overspeeding and pitching techniques for wind turbines, whereas this one reviews the literature on control techniques for enabling wind turbines to participate in tasks related to frequency control. The studies introduced cover several aspects, such as various methods to perform the required speed and pitch angle regulation for wind turbines, the combination of both techniques for various wind speed regions, and the coordination of the power reserves of several wind turbines by WPP central controllers.

In Moutis, Papathanassiou, and Hatziargyriou [74], a lookup table containing the curves shown in Figure 3.9 is used to determine the power setpoint that drives the low-level active power control system of the rotor-side converter of the generator while applying overspeeding techniques. This lookup table takes the required power margin and the rotor speed measurement as inputs. No wind speed measurements are needed. The determination of the power setpoint in the above rated wind speed region is carried out by means of a second lookup table, which takes into account the required power margin and the pitch angle measurement. It also takes into account cubic interpolation between close pitch angles, due to their large influence on the aerodynamic behavior of a wind turbine. Finally, it is noteworthy that overspeeding and pitching actions are not carried out at the same time, but overspeeding only applies when the pitch angle is set to zero.

Approaches that take the wind speed measurement as input can also be found in the literature. For instance, de Almeida, Castronuovo, and Peças-Lopes [73] propose to access the optimum power curve using wind speed as input and determine the optimum power reference P_{opt} and its corresponding optimum rotor speed ω_{opt}. From these signals and the required power margin x, one can determine the deloaded optimum generating point. In this way, the accuracy and reliability of the wind speed measurements become key issues for the implementation of the proposed control strategy. However, wind speed measurement on top of the nacelle is not really reliable when it comes to representing the free wind speed, due to the influence of the rotor.

Another approach proposes combining rotor speed and pitch angle regulations at the same time in the below rated rotor speed range [77]. The idea is to control the above-mentioned parameters in order to maintain the required deloaded level of the wind turbine. Thus, when a frequency drop occurs, the wind turbine shifts both the reference pitch angle and the reference rotor speed towards their optimal values (i.e., it shifts the pitch angle to a minimum value and the rotor speed according to the optimal power curve of the wind turbine). It is worth pointing out how the setpoints of the rotor speed and pitch angle are determined in order to provide the required power reserve. Given a required deloaded power level, the deloaded aerodynamic power coefficient $C_{p_{del}}(v_w, \omega_t, \beta)$ can be determined by applying Equation (3.6). For each wind speed, $C_{p_{del}}(v_w, \omega_t, \beta)$ is influenced by the selected combination of ω_t and β, and there is more than one possible combination. This allows a combination with maximized ω_t to be chosen, which means maximizing the kinetic energy stored in the rotor. This is realized by means of an optimization procedure. As this procedure is complex and time-consuming, the results obtained offline were extrapolated to a suitable online strategy. It is worth highlighting that wind speed measurements are needed in order to implement this control algorithm. When a frequency drop occurs, and due to the high initial rotor speed, the speed regulation of the wind turbine sends a large amount of the injected kinetic energy into the grid in the first few seconds of the frequency excursion. This strategy only applies to the below rated rotor speed range. For near or above rated wind speed levels, in order to avoid exceeding the rated rotor speed level, only pitch control is used to realize the turbine power reference. As a result of the maximization of the kinetic energy injected in the first few seconds of the frequency disturbance, the frequency nadir is delayed in time and its value is higher than in the case of applying a deloaded strategy without this kinetic energy support. However, it is worth noting that the fast power regulation means large loads on the shaft of the turbine.

The idea of taking advantage of the kinetic energy stored in the rotating parts of a wind turbine while in deloaded operation is further investigated by Zertek, Verbic, and Pantos (authors of [77]) in a second paper [78]. Here, new control algorithms are presented for both the pitch angle and the rotor speed that allow the maximization of the kinetic energy in the rotor in order to improve the frequency control support. It is worth noting that both articles conclude that using the kinetic energy stored in the wind turbine reduces the need for deloaded operation while still providing the required amount of power in a short time frame, which is considered to be the maximum deployment time of 100% of the primary reserves according to the ENTSO-E's requirements [48]; that is, 30 s (see Table 3.1).

Deloaded operation of wind turbines that also take into account the interaction with wind farm central controllers are examined in Rodriguez-Amenedo, Arnalte, and Burgos [81], Abdelkafi and Krichen [82], and de Almeida, Castronuovo, and Peças-Lopes [73]. It is worth noting that reliable and fast supervisory control and data acquisition (SCADA) systems with a sampling rate of 1–3 s for frequency measurements should be adequate, considering the desired reaction time of wind turbines [72].

In de Almeida, Castronuovo, and Peças-Lopes [73], the dispatch function of a wind farm central controller is based on the solution of an optimization problem, which sets adjustments of the active and reactive power setpoints for each wind turbine so as to optimize the power flow within the wind park; that is, to minimize the active power losses while participating in tasks related to frequency control.

A wind farm active and reactive control system that provides power setpoints for each wind turbine is also considered in Rodriguez-Amenedo, Arnalte, and Burgos [81]. Once the local controller of a wind turbine receives the active power setpoint from the central controller, two situations may arise. It could be that the wind speed is high enough to make it possible to achieve the required generation level. Then, a PI controller is driven by the error between the active power setpoint and the measured power signal. The output of this PI is the required pitch angle. Therefore, the turbine is always operated at the maxima of the pitch angle curves of Figure 3.7, while the PI controller chooses the pitch angle curve. While computing the pitch angle, an MPT algorithm computes a speed reference from the measured power generation. Since the output power is reduced by the pitch controller, the MPT algorithm outputs a speed reference below its optimal value, yielding reduced values of the tip speed ratio and the C_p power coefficients and then, helping the wind turbine to reduce its generation level. On the other hand, if the wind speed is not high enough for the required generation level to be achieved, the pitch angle is saturated at its minimum value, hence maximizing the power extracted from the wind. It can be deduced that deloaded operation can also be carried out using the control techniques presented in this article. Note that this article does not analyze the stability issues derived from the interaction between the PI controller for pitching the turbine and the central control system of the WPP that sets out the active power reference for the wind turbines.

The following paragraphs focus on the description of pitching techniques according to a literature review. A pitching technique for deloaded operation is presented in Holdsworth, Ekanayake, and Jenkins [83]. Pitch regulation is defined for two regions, denoted as A (above rated power output) and B (below rated power output). In region A, the active power output of the turbine is usually limited to its rated value by setting a nonminimum pitch angle. However, the turbine is deloaded in this case through a further pitch angle increment. In this manner, it is possible to ramp the output of the turbine up and down for frequency regulation. In region B, the active power control scheme of the rotor-side converter controller is commonly commanded by a MPT algorithm [63] (considering variable-speed wind turbines). In this region, the pitch angle is usually set to its minimum value in order to extract the maximum power from the wind. However, in this article the pitch angle is also regulated, in response to the measured frequency, to a change in frequency in a narrow band of ± 2 degrees. In detail, a frequency – pitch angle droop characteristic is applied. This narrow pitch angle regulation band deals up to 400 kW of power spill for a 2 MW DFIG-based wind turbine.

In Moutis *et al.* [75], the proposed deloaded technique is applied to DFIG- and PMSG-based wind turbines. As the power extracted from the wind is linearly dependent on the power coefficient C_p, a defined percentage reduction of power generation can be achieved by reducing in the same proportion C_p (see Section 3.2.1.1). The wind turbine is always operated with an optimal tip speed ratio λ_{opt} (and thus with an optimal C_p when no deloaded operation is commanded). This means that a reduction in C_p from its optimal value can be achieved by determining a nonminimum pitch angle β. This pitch angle is mathematically computed using the following relation:

$$C_p(\lambda_{opt}, \beta) = (1 - x)\% \times C_{p,opt}(\lambda_{opt}, \beta_{min}). \tag{3.7}$$

Given known values of the desired load percentage reduction x, λ_{opt}, and $C_{p,opt}$, the desired value of $C_p(\lambda_{opt}, \beta)$ can be determined. Again, with given λ_{opt}, the required pitch angle β can be determined. Below the rated wind speed, the required power margin is achieved by applying the calculated pitch angle. Above the rated wind speed level, this computed pitch angle is added to the conventional PI pitch controller, which is in charge of limiting the rotor speed to its rated level.

Finally, it is important to note that the contribution of offshore WPPs to main grid frequency control support is also investigated. For instance, Haileselassie *et al.* [84] consider an offshore WPP connected to the grid via a voltage source converter–based high voltage DC link (VSC-HVDC). The WPP is composed of full-converter wind turbines. The objective is to regulate the output of the wind farm in response to main grid frequency variations. The power output of the wind turbines is governed by the offshore-side VSC of the HVDC transmission and this converter does not receive any measurement of the main grid frequency. In order to communicate information on the main grid frequency, the VSC terminal on the onshore side ramps the voltage level of the HVDC transmission up and down proportional to the frequency deviation. The results show that the offshore WPP can lend considerable support to the grid frequency control during grid disturbances. The deloading method used for the wind turbines is not discussed.

3.2.2 Synthetic Inertia

As discussed in Section 3.1.5, in order to promote high penetration levels of wind power into the grid without compromising frequency stability, WPPs may in future be required to provide synthetic inertia [62]. Providing this means replicating the behavior of synchronized generating units with respect to power imbalances in the grid. This can be achieved by introducing control schemes that detect grid frequency variation and command according to active power feed-in. Both the dynamics and the dependency on frequency should be similar to the behavior of synchronous generators. However, the details have not been clearly defined as yet. As described in Equation (3.2), the additional electric power fed in by synchronous machines during frequency changes (i.e., generator deceleration) is proportional to the derivative of their mechanical rotational speed ω_g. This means that for wind turbines, the additional active power (or torque) should be proportional to the derivative of the grid electrical frequency df/dt.

According to Morren, Pierik, and de Haan [85] and Keung et al. [86], typical wind turbine inertia constants are between 4 and 6 s. These values are comparable with the normalized inertia constant of conventional generating units (between 2 and 9 s, depending on the type of generator). Using the fact that the rotor speed of variable-speed wind turbines is not coupled to the grid frequency, the deceleration of the rotor can be chosen by the controller. This allows a tradeoff between the additionally provided power and the duration. The generator speed of conventional synchronous generating units varies directly with frequency; that is, for variations between 47.5 and 52.5 Hz, it stays within 0.95–1.05 pu. In contrast, the generator speed of wind turbines can vary down to 0.7 pu. This means that wind turbines can use more than four times the capacity of regulation of the kinetic energy of conventional synchronized generating units [40]. However, a recovery strategy for the proper rotational speed of the wind turbines after their deceleration for synthetic inertia is required.

The initial loading – that is, the initial rotor speed of wind turbines – has a great influence on their provision of synthetic inertia, as the kinetic energy depends on the square of the rotational speed (see Equation (3.3)). Additionally, the ratings of the converters of the turbine [39] have to be taken into account for the evacuation of the kinetic energy through these power electronics.

In normal operation, the torque of a wind turbine is governed by the maximum power tracking (MPT) algorithm, which does not react to changes in system frequency. However, the following paragraphs discuss several control approaches for torque to respond to the system frequency in order to provide synthetic inertia.

In Ramtharan, Ekanayake, and Jenkins [79], a control system for synthetic inertia is proposed. Figure 3.10 depicts its topology. As shown, the upper path (MPT) contains the conventional determination of the generator torque reference as explained in Section 3.1. Moreover, the loop L1 is in charge of additionally providing an offset torque signal proportional to the ROCOF; for example, a positive, decelerating torque signal if the frequency drops. This decelerating torque signal lasts until the frequency stabilizes. Then, without the support of any additional control action, the overall torque reference T^*_{elec} would decrease, as the purpose of the MPT is to lead the system back to the optimal curve. This would obviously reduce the power injected to the grid transiently and thus take back the frequency support provided directly. Even though recovery of the original turbine speed is necessary, the process should be carefully planned. In particular, recovery should happen slowly, with enough time for the primary frequency controllers in the grid to react.

In order to avoid this reacceleration of the turbine, the control system under consideration also includes a second loop, L2. It is worth noting that this is not, however, commonly found in the literature. This is responsible for providing an additional

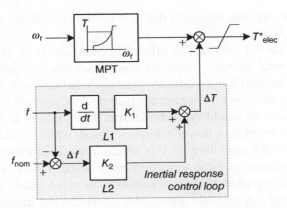

Figure 3.10 The determination of the electromagnetic torque setpoint from an MPT algorithm and an additional control loop for synthetic inertia. *Source:* Adapted from Ramtharan, Ekanayake, and Jenkins, 2007 [79].

torque signal proportional to the frequency deviation Δf, so its output lasts until the nominal frequency level is recovered. Note that this loop L2 is not actually the same as the droop of the primary frequency control, as it provides a torque (not a power reference) depending on Δf.

Conroy and Watson [87] adopt the previously presented control system for synthetic inertia. It can be noted, talking in terms of Figure 3.10, that both K_1 and K_2 are varied with regard to the load level of the wind turbine. This is done because, as discussed in the article, inadequate control parameters can cause unstable operation of the wind turbine. For instance, an excessive value of the proportional parameter K_2 under low wind conditions can cause the wind turbine to stall because of excessive extraction of kinetic energy. The control scheme also considers a recovery strategy for the wind park, to lead the turbines back to their initial operating point. This recovery strategy is based on instructing each wind turbine to switch off its frequency support at a different time.

Two different control methods for synthetic inertia are proposed in Haileselassie *et al.* [84]. The first method is composed of simply the control loop L1 shown in Figure 3.10. However, a low-pass filter is added after obtaining the frequency derivative signal. The aim of this filter is to avoid high rates of change in the torque setpoint obtained due to noise in the frequency measurement. These undesirable excessive torque variations cause mechanical loads in the drivetrain and may also exceed the current limits of the power converters of the wind turbine. The second control method comprises the second loop, L2, shown in Figure 3.10. This article is devoted to comparing the performance of the above-mentioned control methods in the event of a system frequency disturbance. It is concluded that the droop control of loop L2 determines a lower increase in active power than the synthetic inertia provided by loop L1. This results in lower overcurrents and less mechanical stress for the wind turbine. Needless to say, this assertion depends on the considered magnitude of the parameters K_1 and K_2.

A new synthetic inertia approach is offered in Mauricio *et al.* [88]. This method relies on a conventional primary frequency control scheme, but performed in a fast manner. Therefore, the aim is not explicitly to let the wind turbine behave similarly to a conventional generating unit. However, the turbine does provide frequency support in the same time frame as would synthetic inertia. To do that, a droop characteristic (in MW/Hz) is used to obtain a power reference signal, which is added to the output of the conventional MPT algorithm. In this article, a washout filter is applied to the signal Δf (in order to reject the constant component of the signal), so the input of the power – frequency droop characteristic is zero as soon as the steady state is achieved.

It is worth pointing out that this article also considers the interaction of the fast power regulation of wind turbines with the primary frequency controllers of their near conventional generating units. The fast response of wind turbines following a network power imbalance can slow down the response of conventional generators to some extent. This is because the fast additional power injection of wind turbines partly compensates the power network imbalance (affecting the network frequency). However, their support lasts for a few seconds and then conventional generators, which do not track the real magnitude of the power imbalance from the beginning of the disturbance, are needed to act to recover the network balance by full activation of their power reserves. In order to overcome this response delay of conventional generating units, a communication scheme is proposed between them and the wind turbines.

This article also offers a comparison with the previously mentioned synthetic inertia depicted in Ramtharan, Ekanayake, and Jenkins [79]. In regard to this comparison, care should be taken in setting the above-mentioned proportional characteristic K_2 (see Figure 3.10), as higher values of this parameter can affect the oscillatory modes of an interconnected system. On the contrary, the implementation of the washout filter for the frequency in Mauricio *et al.* [88] favors the mitigation of these oscillatory modes.

Finally, in Keung *et al.* [86], an approach for provision of synthetic inertia by a wind farm is proposed. Despite the fact that partial-load conditions are considered for the analysis, the active power output of wind turbines is adjusted by both the pitch angle and the power converters. In the event of a derivative of frequency, the park controller orders an increment in the active power setpoints to the local controller of the power converters of each wind turbine. Three seconds after the activation of this power increase, a slow recovery process to the initial state of the wind turbines is performed. This slow recovery process is carried out by delaying the particular recovery process of each individual wind turbine. Accordingly, each wind turbine injects an increased level of power for a specific time frame. This strategy smooths the net power injected by the wind farm during the recovery process, which can last for up to 30 s. It is worth noting that no aggregated model of the wind farm is used in order to allow an individual recovery process for each wind turbine and also to take advantage of the spatial smoothing effect of the wind farm.

3.3   Conclusions

The following conclusions can be drawn from this chapter:

- The current grid codes of islanded European networks such as those of the UK and Ireland already consider the participation of WPPs in primary frequency control. In this sense, they require wind turbines to be operated such that they do not extract the maximum available power from the wind. They have to be operated in a deloaded mode instead, in order to be able to ramp their output up and down in the event of a frequency deviation.
- On the other hand, other regulations of the strong European electrical grids, such as the German grid, do not consider the operation of wind turbines in a deloaded mode in normal operating conditions. For instance, the German Grid Code only requires that wind turbines reduce their power injection in the particular case of overfrequency.
- Future trends for European regulations, such as the recent European Network Code developed by the ENTSO-E, indicate the need for deloading of wind turbines to participate in primary frequency control. Therefore, it is concluded that in the near future the provision of power reserves by wind turbines for their participation in tasks related to frequency control will be required not only for islanded grids, but also for the strong continental grids.
- Currently, the grid codes of the islanded European networks require that wind turbines be derated by up to 20% of the theoretically available power, and for periods of time up to 30 min, in the event of a negative frequency deviation. When the frequency level is within the normal operating limits, the wind turbines have to maintain a deloaded level of up to 10% of the maximum available power.
- In the literature, two major methods for deloading wind turbines can be found: pitching methods and overspeeding methods. Both methods are based on the idea of achieving a suboptimal working point with respect to power extraction from the wind.
- Each of the above-mentioned strategies fits best with a different wind speed level. For low and medium wind speed levels, considering that the rated rotor speed is not achieved, overspeeding techniques are preferable. Pitching techniques are best suited considering that the rated rotor speed is achieved.
- Among the articles consulted in the review of literature, approaches can be found that consider the measurement of the wind speed for developing overspeeding techniques. As this measurement is typically not reliable, its application is considered problematic. Moreover, when applying overspeeding and pitching techniques, additional mechanical stresses have to be considered for turbine components, such as the generator shaft and the pitch actuators.

4

Energy Storage Technologies

4.1 Introduction

Electrical energy can be converted to many different forms for storage [89]:

- as gravitational potential energy with water reservoirs
- as compressed air
- as electrochemical energy in batteries and flow batteries
- as chemical energy in fuel cells
- as kinetic energy in flywheels
- as magnetic fields in inductors
- as electric fields in capacitors.

In this chapter, we review several available energy storage technologies that can be used in electric power systems. Among other aspects, the operating principles, the main components, and the most relevant characteristics of each technology are detailed. In order to obtain an overview of the main characteristics of the energy storage technologies presented in this chapter and the differences between them in a comprehensive manner, some tables and graphics, based on the data collected from several publications and manufacturers, are shown. For the sake of completeness, the chapter tangentially addresses other storage technologies, such as thermal storage through molten salts and the "power-to-gas" concept. Figure 4.1 graphically depicts the storage classification presented above.

Finally, the last part of the chapter discusses power conversion systems commonly utilized for the grid connection and control of those forms of storage installations that are not synchronized with the network, such as flywheels, super-capacitors, superconducting magnetic energy storage (SMES), batteries, and flow batteries.

Energy Storage in Power Systems, First Edition. Francisco Díaz-González, Andreas Sumper and Oriol Gomis-Bellmunt.
© 2016 John Wiley & Sons, Ltd. Published 2016 by John Wiley & Sons, Ltd.

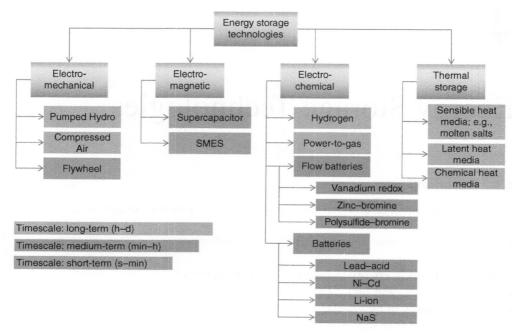

Figure 4.1 The catalog of storage technologies.

4.2 The Description of the Technology

This section covers the description of several storage technologies that can be used for stationary applications. The technologies described include energy storage in pumped hydroelectric installations, such as compressed air, in conventional and flow batteries, by hydrogen-based technologies, in flywheels, in superconducting magnetic devices, and in supercapacitors. These descriptions are mostly based on our previous work. For further details, see Díaz-González *et al.* [42].

4.2.1 *Pumped Hydroelectric Storage (PHS)*

PHS is the most mature storage system amongst those considered in this work. In fact, it is the most used technology for high-power applications [90]. Its operating principle is based on managing the gravitational potential energy of water, by pumping it from a lower reservoir to an upper reservoir while consuming power from the grid, or by releasing water from the upper reservoir to the lower one when energy needs to be injected into the grid (Figure 4.2). The energy stored is proportional to the volume of water in the upper reservoir and the height of the waterfall. So,

$$E_{PHS} = \rho g H V, \tag{4.1}$$

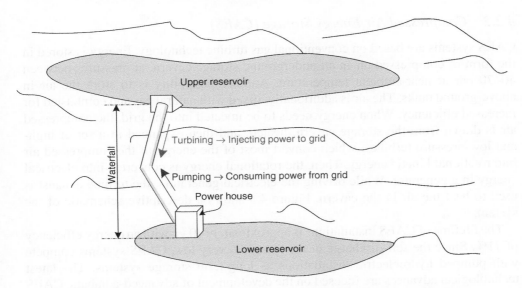

Figure 4.2   The operating principle of pumped hydroelectric storage (PHS).

where E_{PHS} is the stored energy, in joules, ρ is the density of the water, equivalent to 1000 kg/m^3, g is the acceleration due to gravity, equivalent to 9.8 m/s^2, H is the height of the waterfall (in meters), and V is the volume of water stored in the upper reservoir.

There is a huge global hydro-storage potential nowadays, estimated at approximately 3000 GW [91]. In the European Union (EU) there is about 7400 MW of new PHS installations projected, which is a 20% increase in the EU's installed capacity [92].

The technology used in pumped hydroelectric installations includes highly developed reliable devices such as electric generators and hydraulic turbines that allow the system to regulate vast power levels with high ramp power rates. For example, an installation by the First Hydro Company [56], commissioned in 1984, is capable of moving from 0 to 1320 MW power injection in only 12 s. The lifetime of pumped hydroelectric installations is around 30–50 yr, with an acceptable round-trip efficiency of 65–75% and power capital costs of (€ 500–1500)/kW and (€ 10–20)/kWh [93]. This cost estimation can be greatly affected by the fact that the construction of a pumped hydroelectric system commonly depends on the availability of a very hilly geographical environment, as at least 100 meters of waterfall between the upper and lower water containments is needed. The preservation of the natural environment and the possible absence of electric transmission lines in places with suitable topography for the installation of pumped hydroelectric systems must be taken also into account in their commissioning.

4.2.2 Compressed Air Energy Storage (CAES)

CAES systems are based on conventional gas turbine technology. Energy is stored in the form of compressed air in an underground storage cavern, at pressures between 40–70 bar at near-ambient temperature. Another possibility is to store the air in above-ground tanks. The air is additionally mixed with natural gas and combusted for increased efficiency. When energy needs to be injected into the grid, the compressed air is drawn from the storage cavern, heated, and then expanded in a set of high- and low-pressure turbines, which convert most of the energy of the compressed air into rotational kinetic energy. Then, the rotational energy is converted into electrical energy in a generator. While driving the electrical generators, the turbine exhaust is used to heat the air in the cavern. Figure 4.3 shows a descriptive schematic of the system.

The lifetime of CAES installations is approximately 40 yr, with an energy efficiency of 71%. Since the self-discharge of the system is very low, CAES systems compete with pumped hydroelectric installations as long-term storage systems. The latest technological advances are focused on the development of advanced-adiabatic CAES (AA-CAES). In this system, the air is adiabatically compressed and then pumped into an underground cavern. The effectiveness and the economics of the required heat exchangers for the adiabatic compression, the compressor, and the expander trains are the main concerns for the success of AA-CAES.

The availability of suitable underground caverns is a key aspect for the installation of CAES systems. As an example, the Iowa Stored Energy Park [94], which was considering the installation of a 270 MW CAES system for storing wind power during periods of low demand, recently cancelled the project for this reason. In particular, geological studies of the site showed that the storage reservoir was not suitable for the scale of the project. There are just two CAES installations in the world so far, one

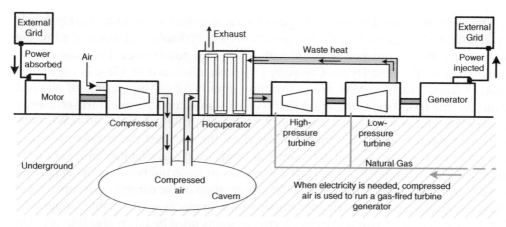

Figure 4.3 The operating principle of compressed air energy storage (CAES).

in Germany, with a rated power of 290 MW and the other in Alabama, United States, with a rated power of 110 MW.

4.2.3   Conventional Batteries and Flow Batteries

4.2.3.1   Basic Concepts

A battery energy storage system (BESS) converts electrical energy into potential chemical energy while charging, and releases electrical energy from chemical energy while discharging. In general terms, it is based on reduction and oxidation reactions (commonly called redox reactions). An electrochemical reduction reaction is one that allows the component involved to gain electrons, while an oxidation reaction allows the component to lose electrons. Redox reactions yield new active electrochemically active substances with nonneutral electric charge, and hence ions.

A battery cell is just a device that provides the conditions for redox reactions to happen, thus generating a flow of ions and electrons between the areas in which these occur. The flow of electrons and ions exists as long as there is an energy difference between the electrochemically active substances involved in the reduction and oxidation reactions. To enable this flow of ions and electrons, the battery cell has two circuits, one external and the other internal. The internal circuit is comprised of the battery cell itself, and provides the path through which the resultant ions flow. The electrical circuit is closed by adding the external circuit, thus providing the path through which the electrons resulting from the redox reactions can flow. This external path is provided by the external system (either a load or an energy source) to which the battery is connected. The battery cell is comprised of the following components:

- **The electrodes**. While discharged, oxidation reactions occur in the anode of the battery (the negative electrode), which is the electrode that captures the electrons lost by the component. Conversely, reduction reactions occur in the cathode of the battery (the positive electrode), which is the electrode that provides the electrons gained by the reduced component.
- **Two pairs of electrochemically active substances**. There is one in the anolyte region while the other is in the catholyte region. The materials composing the anolyte electrode and the component or substance surrounding it have to react, yielding an oxidation reaction (while discharged). Analogously, the electrochemical interaction between the materials comprising the catholyte electrode and the substance or component surrounding it yields a reduction reaction. The above defines the two pairs of electrochemically active substances.
- **The electrolyte**. Apart from causing the two pairs of electrochemically active substances to gain or lose electrons, the redox reactions yield ions (and hence particles with a nonneutral electric charge). To ensure the equilibrium of charge

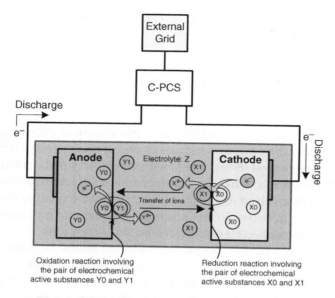

Figure 4.4   The operating principle of a battery.

between the anolyte and catholyte electrochemically active substances, these ions are exchanged between them. This ion transport is enabled by the electrolyte, which is a solid or liquid electronically insulating substance.

- **The separator**. There is an electrical potential between the electrochemically active substances in the anolyte and catholyte regions. The separator avoids direct contact between them, thus preventing the battery from an internal short circuit.
- **The container**. Batteries are composed of several cells, either in series or in parallel to achieve the desired electrical characteristics. In the container, they are all packed into a controlled and isolated environment.

Having presented the main components of a battery cell, now it is much easier to understand the operating principle, and this is graphically supported by Figure 4.4. As can be noted, the battery cell is composed of two electrodes, made up of two materials called Y0 (for the anode) and X0 (for the cathode). They are both surrounded by the electrolyte, Z. Also, the anode (negative electrode) is surrounded by the substance or component Y1. Similarly, the cathode is surrounded by the component X1. The materials X0–X1 and Y0–Y1 define two pairs of electrochemically active substances. The difference in energy state of the two pairs of electrochemically active substances is translated into a voltage difference. While charged, the voltage between the electrodes of the cell is at a maximum and varies between 1 and 4 volts, depending on the technology used, yielding the so-called open-circuit voltage of the cell.

By adding an external load between the electrodes, the electrical circuit is closed. Then, the battery is being discharged, and this means that redox reactions start to occur, yielding an electric current through the load. The electrons flow from the negative electrode (the anode, the region with the maximum energy state) to the positive one (the cathode, the region with the minimum energy state). These electrons, and the positive ions Y^{2+}, are the result of the oxidation reaction between substances Y0 and Y1. The electrons are collected by the catholyte electrode, yielding a reduction reaction between substances X0 and X1, which in turn results in the ion X^{2-}. The internal circuit allows the ionic exchange. The ion(s) exchanged depend on the technology used and are usually those presenting the highest mobility. As a result of this process, each of the pairs of electrochemically active substances is weakened, so the electric potential between them is diminished. The electrical potential between the two electrodes can be restored by reversing the flow of the electric current, and hence by applying an external energy source to charge the battery.

The electric potential derived from the chemical reactions by the two pairs of electrochemically active substances, measured in the full charge state of the cell when disconnected from any circuit, is called the *open circuit voltage* V_0. This voltage determines the free energy of reaction (the Gibbs free energy) of the electrochemical reactions in the battery cell.

The open-circuit voltage can be measured with a voltmeter, and it corresponds to equilibrium conditions. This means that the open-circuit voltage cannot be measured immediately after a battery charging process is completed but only a few hours later, thus permitting the battery to "relax" after the process (this time depends on the battery type and the method of charging). Another method to estimate the open-circuit voltage is a statistical one, based on evaluating several voltage curves during the charging process using a decreasing sequence of currents, as explained in Snihir, Rey, and Verbitskiy [95].

It is quite important to differentiate concept of the the open-circuit voltage from the voltage measured in the cell during the discharging process, as commonly presented in manufacturers' datasheets. These graphs show the voltage profile during discharge at determined current rates (see Figure 4.5).

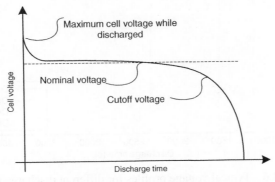

Figure 4.5 A typical voltage–discharge profile for a battery cell.

Due to the internal resistance of the cell and the electric current flowing through it, there is a voltage drop in the cell, so one should not consider the cell voltage measured at the beginning of the discharging process (and hence under full charge conditions) to be equal to the open-circuit voltage. By discharging the cell, the voltage decays from its maximum value when fully charged to the so-called *cutoff voltage*, which defines the usable voltage range of the cell. From this point on, the voltage decays dramatically, strongly limiting the usability of the device.

Manufacturers usually indicate a "nominal cell voltage." This is just an averaged value between the maximum and the cutoff voltage. This average cell voltage, though, in conjunction with the discharge current rate, serves for the calculation of the equivalent internal resistance of the cell.

The discharge voltage profile of the cell depends on many factors, such as pressure and temperature, since they affect the performance of the chemical reactions in the cell. The cell voltage and the discharge current rate are somewhat related, thus affecting the energy capacity. This is why manufacturers are supposed to indicate the capacity of the cells for different discharge current rates, also indicating the applied control discharge method [96]. Figure 4.6 shows typical voltage trends for battery cells at

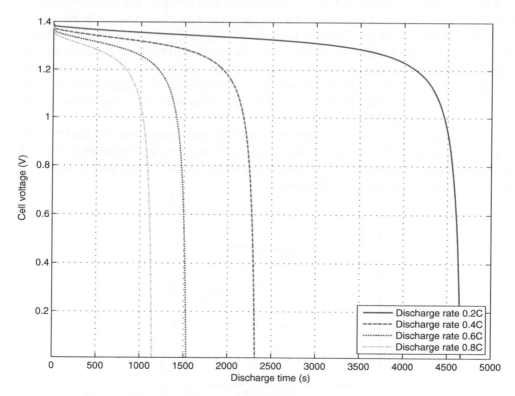

Figure 4.6 Typical voltage profiles for different discharge rates.

different current discharge rates. As can be noted, the higher the current discharge rate, the lower are the maximum and the cutoff voltage of the cell, thus decreasing the energy storage capacity.

The *energy capacity* of batteries is defined as the usable energy at a defined discharge rate. This term is commonly indicated in amperes per hour (Ah). It is important to understand the specific nomenclature commonly used while defining the energy capacity. For example, we could read in datasheets that a battery is rated at 10 Ah at 2 C, or 10 Ah at C/5. In the second case, the battery will provide 10 Ah/5 h = 2 A over 5 h, until reaching its cutoff voltage. Similarly, in the first case, the battery will provide 10 Ah/0.5 h = 20 A for 30 min.

The ratio between the remaining energy capacity and the rated capacity of the battery (at a defined discharge rate) determines the state of charge (SoC) of the system. The charging and discharging processes are not ideal, but they account for losses that are translated into voltage drops and temperature rises. Thus, the energy efficiency can be formulated in terms of the cell voltage and the theoretical maximum voltage produced in the chemical reaction in the cell; the so-called *thermodynamic voltage*. While charging, the energy efficiency in the cell is defined as

$$\mu_c = V/V_{th}, \tag{4.2}$$

and during discharging, it can be formulated by

$$\mu_d = V_{th}/V, \tag{4.3}$$

where V_{th} is the thermodynamic voltage and V is the voltage between the cell terminals. These metrics are formulated differently since the voltage in the cell is higher than the thermodynamic one in discharge, and lower during the charging process. This is because of the effect of the internal resistance of the cell (for further details on electrical battery models, see Chapter 6).

With regard to the definition of the energy efficiency, and the dependency of the voltage cell on the SoC (see Figure 4.5), it can be derived that the discharging efficiency increases with the SoC. Looking at this the other way round, the charging efficiency decays with the SoC. Therefore, to maximize the overall efficiency throughout a full charge and discharge loop, an optimal average SoC should be calculated. This duty is usually assigned to battery management systems (see Section 4.3.4 for further details).

Further principal magnitudes for energy storage in general, and for batteries in particular, are the specific energy and the power. These magnitudes relate the energy and power capacity to the battery weight. Only cell reactants are involved in the calculation of these metrics, thus leaving out other structural components, cell separators, and wiring, for instance. The energy capacity can be calculated as follows [97]:

$$E_{sp}(Wh/kg) = \frac{nFV_{th}}{3.6 \sum M_i},$$ (4.4)

where n is the number of electrons transferred in a chemical reaction, F is the Faraday constant in coulombs per mole, V_{th} is the theoretical thermodynamic voltage of the cell, and $\sum M_i$ is the sum of the molecular weights of the reactants of the cell.

Thus, to optimize the specific energy, one should maximize V_{th} and minimize $\sum M_i$. The latter is achieved by configuring cells based on reactants with low atomic weights. The former is obtained by building up the two pairs of electrochemically active substances with a highly electropositive element and a highly electronegative element. When that is done, the chemical reactions will yield a high thermodynamic potential V_{th}.

Finally, the specific power determines the maximum power that the cell can deliver in relation to its weight. According to Ehsani, Gao, and Emadi [97], it depends on the open-circuit voltage of the cell, V_0, and an internal resistance R_{int}, apart from the ohmic resistance of conductors, R_c. Thus

$$P_{sp}(W/kg) = \frac{V_0}{4(R_c + R_{int})}.$$ (4.5)

This variable internal resistance, usually called the overpotential, models a voltage drop in the cell and depends on the amount of current drawn. Chapter 6 provides an in-depth treatment of the modeling of battery cells.

Looking at the electrochemistry, and with the aim of maximizing the specific energy, one can come up with several types of battery cells. As previously noted, the electrochemically active substances used for building up battery cells provide different open-circuit voltages. Other characteristics – such as, for instance, the cyclability, the time response, the operational temperature range, and ageing – vary considerably with the technology used. That is why research and development (R&D) is so intensive in this field.

The oldest battery type is the lead–acid one, for which R&D activity has been carried out for more than 140 yr. But there are other several technologies that are gaining importance in different stationary and nonstationary fields of application. These are, amongst the most popular types, the nickel–cadmium (Ni–Cd), sodium–sulfur, and lithium-ion (Li-ion) types, and the so-called flow batteries. The principal specificities of each technology are presented in the following sections.

4.2.3.2 Lead–Acid Batteries

Research involving lead–acid batteries has been conducted for over 140 yr. There are two major types of lead–acid battery: flooded batteries, which is the most common topology, and valve-regulated batteries, which are the subject of extensive R&D.

Commonly, lead–acid battery cells are built up of several lead plates arranged in parallel. These are alternatively polarized, so that the cathodic plates are coated with lead dioxide PbO_2 and the anodic plates with porous lead Pb. The plates are immersed in the electrolyte, which is made up of sulfuric acid H_2SO_4.

The global oxidation and reduction reactions in the cell can be summarized as follows:

$$Pb + SO_4^{2-} \Leftrightarrow PbSO_4 + 2e^- \qquad \text{(anode)},$$

$$PbO_2 + 4H^+ + SO_4^{2-} + 2e^- \Leftrightarrow PbSO_4 + 2H_2O \qquad \text{(cathode)},$$

$$Pb + PbO_2 + 2H_2SO_4 \Leftrightarrow 2PbSO_4 + 2H_2O \qquad \text{(total)}.$$

During the discharging process, the porous lead anode reacts with the sulfuric acid, yielding lead sulfate $PbSO_4$ and an excess of electrons, which are transmitted through the external circuit of the cell (the connected external load) to the cathode. These electrons, among with the sulfuric acid, react with the lead dioxide to also form lead sulfate $PbSO_4$. In addition, water is formed in this process. Since the electrolyte is consumed in the reactions, the specific gravity serves as a guide for estimating the SoC of the battery. The electrical potential between the two electrodes of the cell due to the reactions described above results in around 2.04 V.

Lead–acid batteries suffer from some problems as a result of the way in which they are charged and discharged, one of which is the so-called sulfation. This occurs when the battery is deprived of periodic full-charge processes. In this case, large lead sulfate crystals are formed, which cannot be reversed in the porous lead and lead dioxide in the electrodes of the battery, thus decreasing the battery's capacity. Sulfation is also exacerbated by exhausting the energy stored in the battery, so very deep discharges are not recommended. Another common problem arises when the applied charging voltage surpasses the admissible or recommended level. In this case, the water in the electrolyte can be exhausted by forming hydrogen gas, with the consequent risk of explosion due to its high flammability.

One of the main differences between flooded and valve-regulated lead–acid batteries lies in the way they manage the gases formed in the chemical reactions in the cells. In flooded batteries, hydrogen gas is naturally evacuated from the cell, and the electrolyte can easily be replaced. Valve-regulated batteries are sealed, so these operations are not possible. However, they ensure no electrolyte and gas leakages, thus accounting for any need for a mechanism to recombine the hydrogen formed into water dissolved in the electrolyte.

Lead–acid batteries present the poorest cycle life, just 200–1800 cycles depending on the depth of discharge (DoD) and the operating temperature, among the different types of batteries considered in this work. In addition, it is worth noting the need for periodic water maintenance (of flooded batteries) and the low energy and power densities of this type of battery. In spite of the above-mentioned drawbacks, their use is widespread in both stationary and nonstationary applications. One of the most important advantages of lead–acid batteries is their low cost (up to € 270/kWh [93]) compared to the costs of other types of batteries.

4.2.3.3 Nickel–Cadmium Batteries

Nickel–cadmium (Ni–Cd) batteries are used in both portable and general stationary industrial applications. In portable applications the battery is in its sealed form, and in industrial applications it is in flooded form. It is worth noting that Ni–Cd batteries compete with NiMH in the field of alkaline batteries.

Ni–Cd batteries are primarily produced using nickel and cadmium hydroxide. These materials are then polarized into nickel oxyhydroxide $NiO(OH)$ cathodic plates, and anodic plates of porous cadmium. The battery cells are immersed in an electrolytic aqueous alkaline solution based on potassium hydroxide KOH [98].

To prevent short circuits between adjacent anodic and cathodic plates in the cell, a separator is used. This is generally based on polystyrene or polypropylene, but other options are suitable, such as fibrous polyamide. The selection of the separator is critical, as it can constrain the easy mobility of ions produced in chemical reactions between the electrodes, and this is translated into an increment in the internal resistance of the cell [98].

The global oxidation and reduction reactions in the cell can be summarized as follows [99]:

$$Cd + 2OH^- \Leftrightarrow Cd(OH)_2 + 2e^- \qquad \text{(anode)},$$

$$2NiOOH + 2H_2O + 2e^- \Leftrightarrow 2Ni(OH)_2 + 2OH^- \qquad \text{(cathode)},$$

$$2NiOOH + Cd + 2H_2O \Leftrightarrow 2Ni(OH)_2 + Cd(OH)_2 \qquad \text{(total)}.$$

During the discharging process, the porous cadmium Cd in the anode reacts with the ion OH^-, yielding $Cd(OH)_2$ and the electrons that are transmitted through the external circuit of the cell to the cathode. These electrons, along with the water in the electrolyte, react with $NiO(OH)$ to form $Ni(OH)_2$ and the ion OH^-. The electrical potential between the two electrodes of the cell due to the reactions described above results in around 1.2 V, and hence in a lower potential than in lead–acid batteries. It is interesting to note that the electrolyte KOH is not consumed, so its density (or specific gravity) in the aqueous electrolytic solution does not serve as a measurement of the SoC of the cell.

Ni–Cd batteries present good characteristics with regard to cyclability (more than 3500 cycles [98], and even 50 000 cycles at 10% of DoD), high ramp power rates, and low maintenance. On the other hand, they present three major drawbacks that limit their commercial success. First, the cost of Ni–Cd batteries is very high compared to the cost of lead–acid batteries (more than ten times). Second, cadmium and nickel are toxic heavy metals that can cause health risks in humans. For this reason, the European Commission proposed recycling targets of at least 75% for this type of battery in November 2003. Third, Ni–Cd batteries suffer from the memory effect. When a Ni–Cd battery is repeatedly recharged before becoming fully discharged, a sudden voltage drop in the cell is experienced. According to Broussely and Pistoia [98], this voltage drop is not a sign of a real capacity fade in the cell. However, since the resultant voltage at this point is lower than that needed for the proper management of the associated power conversion, the memory effect is actually regarded as a capacity fade. Furthermore, the memory effect is experienced in sealed Ni–Cd batteries, but not in flooded ones.

4.2.3.4 Sodium–Sulfur Batteries

Sodium–sulfur (NaS) batteries are one of the most promising technologies for stationary high-power applications. The low atomic weight of sodium and the high thermodynamic voltage configure sodium-based battery cells with a high specific power. The cell construction is rather different from those presented so far. The electrodes of NaS battery cells are liquid, while the electrolyte, which in turn acts as a separator, is solid. The negative electrode (liquid sodium) is surrounded by the electrolyte, which is shaped in a tube fashion (see Figure 4.7). The material of the electrolyte is ceramic beta-alumina, and the material for the positive electrode is liquid sulfur (usually embedded in a carbon felt). To bring the electrodes to their liquid state, they have to be melted, imposing operating temperatures for NaS batteries of around 300–400 °C. At this temperature, the ceramic electrolyte is a good conductor of the ions produced in the chemical reactions inside the cell.

The global oxidation and reduction reactions in the cell can be summarized as follows [101]:

$$2Na \Leftrightarrow 2Na^+ + 2e^- \qquad \text{(anode)},$$
$$xS + 2e^- \Leftrightarrow S_x^{2-} \qquad \text{(cathode)},$$
$$2Na + xS \Leftrightarrow Na_2S_x \qquad \text{(total)}.$$

During the discharging process, the sodium in the anode is oxidized into sodium ions, which flow through the electrolyte to the cathode. Then, they combine with

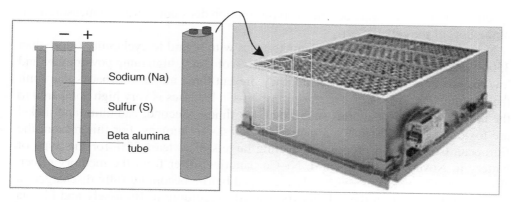

Figure 4.7   A schematic of a NaS battery cell and module. *Source:* The cell graph on the left is adapted from NGK Insulators, Ltd, http://www.ngk.co.jp/english/ (accessed April 22, 2015) [100]; and the photograph of the battery module on the right is courtesy of Wen, Z., Cao, J., Gu, Z., *et al.* (2008) Research on sodium sulfur battery for energy storage. *Solid State Ionics*, **179**, 1697–1701 [101]. Reproduced with permission of Elsevier.

reduced sulfur anions to generate sodium polysulfide, all of this yielding the above cell reaction. The electrical potential between the two electrodes of the cell due to the reactions described above (the open-circuit voltage) results in around 2.075 V at 350 °C.

Sodium–sulfur batteries are relatively recent technology. Among other characteristics, the major advantages are related to their high energy efficiency (85%) and their energy density (151 kWh/m^3) [102], in conjunction with very low self-discharge rates, low maintenance, and almost 99% recyclability. An important feature of these batteries is their high operating temperature, around 350 °C [98]. In this regard, it is convenient to note that since the cell reactions are exothermic, the energy input needed to maintain a proper operating temperature is low and, therefore, the efficiency of the battery is not substantially reduced. These characteristics, in conjunction with a relatively low capital cost (comparable with the cost of conventional lead–acid batteries), define the NaS battery as one of the most promising technologies for high-power storage applications. Amongst the main drawbacks of this young technology, we can highlight the continuing need for intensive R&D to overcome, for example, problems such as cracking of the ceramic electrolytic tube and corrosion due to sulfur, both of which reduce the lifetime of the battery [103].

4.2.3.5   Lithium-Based Batteries

Lithium-ion batteries are currently attracting much attention and are viewed as promising solutions in the field of buildings, electromobility, and renewable generation. For instance, batteries from Tesla Motors [104] are considered to be quite attractive for multiple applications, even taking into account that the manufacturer has yet to start rolling out product deliveries at the time of writing this book. The active material in the cathode (positive electrode) of Li-ion cells is usually lithium metal oxide, in the form of lithium cobalate ($LiCoO_2$). The negative electrode is mainly carbon (C) and lithium atoms are actually in the electrode. The electrolyte is an organic solution containing lithium-based dissolved salts, such as $LiClO_4$ and $LiPF_6$ [105]. Finally, the electrode areas are separated by porous separators based on polyethylene or polypropylene. A schematic of an Li-ion battery is shown in Figure 4.8.

The global oxidation and reduction reactions in the cell can be summarized as follows [106]:

$$Li(C) \Leftrightarrow Li^+ + e^- \qquad \text{(anode)},$$

$$Li^+ + e^- + CoO_2 \Leftrightarrow LiCoO_2 \qquad \text{(cathode)},$$

$$Li(C) + CoO_2 \Leftrightarrow LiCoO_2 \qquad \text{(total)}.$$

During the charging process, lithium ions Li^+ are extracted from the cathode of the cell and get inserted into the graphene sheets in the graphite (the negative electrode). As always, the electrons resulting from the chemical reactions flow, in this case, from the positive electrode to the negative electrode through an external energy source, thus closing the electric circuit. The electrical potential between the two electrodes of the cell due to the reactions described above (the open-circuit voltage) reaches up to 3.7 V.

This high open-circuit cell voltage and the low weight of lithium yield a very high specific energy (around 75–125 Wh/kg) [105]. The energy density also proves to be very high, at around 170–300 Wh/l, and thus Li-ion batteries are well suited for portable applications, such as mobile phones and electronic devices. Other noticeable features are their fast charge and discharge capability and the relatively high round-trip efficiency of 78%. Among the drawbacks of the technology, it is important to note the required narrow voltage and temperature ranges for proper operation, which motivate the need for protection circuits [107]. In addition, the use of flammable organic electrolytes raises issues about security and environmental issues. Nowadays, great efforts are being made in the field of material technology in order to make the Li-ion battery suitable for high-power stationary applications and use in the field of the electromobility.

4.2.3.6 The Flow Battery Energy Storage System (FBESS)

As in the case of conventional batteries, the operating principle of flow batteries (FBESS) is based on the electrochemical reactions that occur in electrochemical cells. However, flow batteries differ from conventional ones in the fact that the electrolyte is not permanently stored in the cells but, instead, two aqueous electrolytic solutions

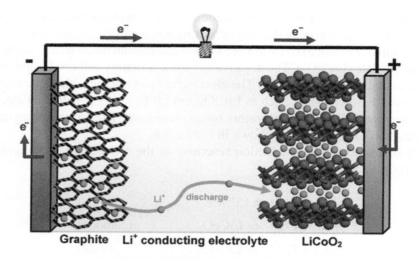

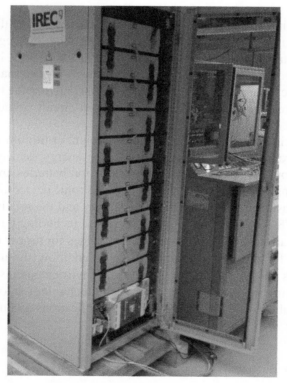

Figure 4.8 A schematic of an Li-ion battery cell (top) and a Saft Li-ion battery pack in the IREC laboratory (bottom). *Source:* Scheme courtesy of Bruce, P.G. (2008) Energy storage beyond the horizon: rechargeable lithium batteries. *Solid State Ionics*, **179**, 752–759 [106]. Reproduced with permission of Elsevier. Bottom photograph courtesy of IREC, Catalonia Institute for Energy Research, http://www.irec.cat/ (accessed April 22, 2015) [108].

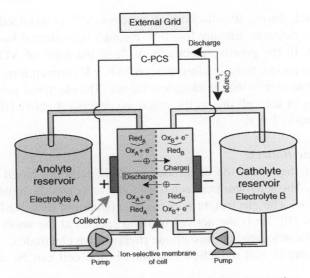

Figure 4.9   The operating principle of flow batteries.

(A and B) are contained in separate tanks. During the charging process, these aqueous solutions are pumped through the electrochemical cells, where the electrochemically active material dissolved in electrolyte A is oxidized at the anode, and the electrochemically active material in electrolyte B is reduced at the cathode. The discharge cycle comprises the reverse process [109]. Figure 4.9 depicts the operating principle of flow batteries.

This section concerns three main types of flow batteries: the vanadium redox battery (VRB), the zinc–bromine battery (ZBB), and the polysulfide bromide battery (PSB). The principal materials and chemistry for the above-mentioned flow battery cells are illustrated in the following paragraphs.

The Vanadium Redox Battery

Both the anode and cathode of the cells are composed of catalyzed graphite. The aqueous electrolyte for both the anodic and the cathodic regions is based on sulfuric acid H_2SO_4, in which vanadium sulfates are dissolved as active chemical species. The anodic and cathodic regions are separated by a polymer membrane, which permits ion exchange between the two electrodes.

The global oxidation and reduction reactions in the cell can be summarized as follows [110]:

$$V^{2+} \Leftrightarrow V^{3+} + e^- \qquad \text{(anode)},$$

$$VO_2^+ + 2H^+ + e^- \Leftrightarrow VO^{2+} + H_2O \qquad \text{(cathode)},$$

$$V^{2+} + VO_2^+ + 2H^+ \Leftrightarrow V^{3+} + VO^{2+} + H_2O \qquad \text{(total)}.$$

As can be noted, during the discharging process, V^{2+} is oxidized in the anode, yielding V^{3+}. The electrons that are lost flow through the external load to which the cell is connected. In the positive electrode, V^{5+}, in the form of VO_2^+, accepts the electrons from the anode, thus yielding the ion VO^{2+}. Hydrogen ions are exchanged through the membrane, closing the electrical circuit. The electrical potential between the two electrodes of the cell due to the reactions described above (the open-circuit voltage) reaches up to 1.6 V.

The Zinc–Bromine Battery

As in the case of the VRB, the electrodes of ZBB cells are based on a carbon–plastic composite. The separator between the anodic and cathodic regions is made up of polyolefin sheets. The electrolyte is aqueous, containing dissolved zinc bromide salts (see Figure 4.10). Zn is the active chemical species at the anode (the negative electrode), while Br is located at the cathode (the positive electrode).

The global oxidation and reduction reactions in the cell can be summarized as follows [111]:

$$Zn_{(s)} \Leftrightarrow Zn^{2+}_{(aq)} + 2e^- \qquad \text{(anode)},$$

$$Br_{2(aq)} + 2e^- \Leftrightarrow 2Br^-_{(aq)} \qquad \text{(cathode)},$$

$$Zn_{(s)} + Br_{2(aq)} \Leftrightarrow 2Br^-_{(aq)} + Zn^{2+}_{(aq)} \qquad \text{(total)}.$$

The Zn is dissolved – oxidized from a solid state attached to the surface of the negative electrode – into Zn^{2+} ions. In turn, the bromide ($Br_{2(aq)}$) is converted into

Figure 4.10 The zinc–bromine flow battery shown on the right builds up to the containerized system shown on the left. *Source:* Redflow Limited (2015) [113], http://www.redflow.com.au/ (accessed April 22, 2015). Reproduced with permission of Redflow Limited.

bromine Br⁻, all of which yields the global cell reaction indicated above, providing an electrical potential between the electrodes of around 1.7 V. As can be noted, ideally the Zn and Br should not react directly. If this happens, it will be translated into battery self-discharge. The separator prevents the bromide ions from reaching the negative electrode. Also, the cathodic electrolyte is enriched with organic components of diverse nature [112, 111], with the aim of capturing and forming, with bromide, an emulsion that is insoluble in water, and that sinks to the bottom of the positive electrolyte tank. Reactions that are the reverse of those described above take place during the battery charging process.

The Polysulfide–Bromide Flow Battery

In PSBs, the electrodes are based on a carbon–plastic composite. The separator between the anodic and cathodic regions is made up of polyolefin sheets. The aqueous electrolytes are based on sodium polysulfide Na_2S_y in the anodic region, and sodium bromide NaBr in the cathodic one.

The global oxidation and reduction reactions in the cell can be summarized as follows [114]:

$$2Na_2S_2 \Leftrightarrow Na_2S_4 + 2Na^+ + 2e^- \qquad \text{(anode)},$$

$$NaBr_3 + 2Na^+ + 2e^- \Leftrightarrow 3NaBr \qquad \text{(cathode)},$$

$$2Na_2S_2 + NaBr_3 \Leftrightarrow 3NaBr + Na_2S_4 \qquad \text{(total)}.$$

As can be noted, during the discharging process the sodium polysulfide in the anode is oxidized, losing electrons that serve to reduce the active chemical species in the cathode. The polymeric separator permits the exchange of sodium ions, $2Na^+$. The electric potential between the electrodes of the cell can achieve 1.5 V.

In general terms, flow batteries are easily scalable, since the volume of the stored electrolyte determines the energy capacity of the system. The power capacity depends on the number and size of the electrochemical cells [115]. Also, they are suitable for storing energy over long periods of time due to their very low – and even negligible – self-discharge.

Flow batteries can be fully discharged without any damage, and in terms of cyclability, they present better characteristics than conventional batteries. For instance, the VRB can achieve 13 000 charge and discharge cycles at 100% of DoD, with a relatively high energy efficiency of 78%.

Turning now to the economics, the cost of these batteries is comparable to that of NaS batteries (around 500 US$/kWh in the case of the ZBB, for instance [116]). In fact, in general they are cheaper than Li-ion or Ni–Cd batteries.

Another common characteristic of these types of batteries is the fact that they require very low maintenance, specially in the case of the VRB, as it uses the same electrolyte in the anode and the cathode, avoiding the risk of cross-contamination of the aqueous solutions. Electrolyte cross-contamination exists in PSBs, though, because of the exchange of sodium ions. In terms of greenness, flow batteries are basically made of recycled plastics, allowing low-cost production and high recyclability. However, in the case of the PSB it is worth noting that toxic bromine gas would be expelled in the event of tank failure [115].

The operating temperatures of the reviewed flow batteries are in the range of 10 to 40–45 °C [113]. Without any doubt, the major limitation of the technology is its relatively low specific energy (in Wh/kg), which restricts its use to stationary applications. In this regard, it is worth noting that the specific energy for the ZBB is logically higher than for the VRB. Indicative figures are around 25–35 Wh/kg for the VRB [112] and 70–90 Wh/kg for the ZBB [117].

Finally, we should note that this is not a mature technology yet, especially in the case of the PSB. In fact, amongst the three types of flow batteries reviewed in this chapter, the PSBs are the least developed, with just a couple of pilot plants worldwide. In 2003, Regenesys Technologies built a pilot plant in South Wales, rated at 15 MW/120 MWh. However, with a project budget of around $250 million, the system was never fully commissioned. VRB Power Systems, Inc., founded in 2004, was acquired by Prudent Energy in 2009 [118]: currently, no specific reference to PSB technology can be found in the company's product brochure.

4.2.4 The Hydrogen-Based Energy Storage System (HESS)

The most common option for the production of hydrogen is from coal or other fossil fuels [119]. However, it can also be obtained by means of water electrolysis, from various forms of renewable energy, and from gasifying biomass. Once the hydrogen has been produced, it can be transported through pipelines to the users to produce electricity, or stored in order to be used later in fuel cells. The so-called regenerative fuel cell (RFC) [117] comprises the production process of hydrogen using a water electrolyzer, a hydrogen storage medium, and a fuel cell system, which allows the production of electricity from the stored hydrogen. This is the considered topology in this work for the HESS. In this case, the electric power from wind facilities could be used to feed the electrolyzer to produce the hydrogen. Figure 4.11 depicts the presented concept.

Some of the principal elements of RFCs that are attracting much attention nowadays with regard to R&D are the electrolyzer and the fuel cell, since their performance is critical in order to maximize the overall energy efficiency. As previously noted, the electrolyzer, by using an electricity source, converts water (e.g., 1 mole) into hydrogen (1 mole) and oxygen gas (a half-mole). Then, this hydrogen serves as a "'fuel" in the fuel cell, in conjunction with oxygen gas, to produce electricity, water, and heat, as products of the electrochemical reaction in the cell. Thus, the electrolyzer and the fuel cell perform the same reactions but in opposite directions. This is why some fuel cell

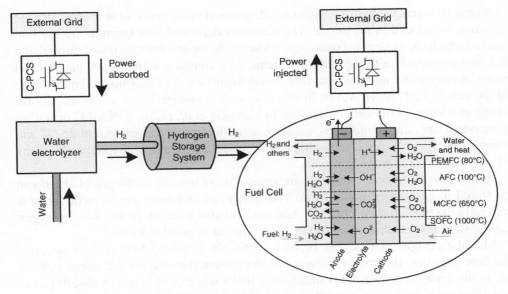

Figure 4.11 The concept of the RFC.

technologies are also used as electrolyzers by reversing the electric current flowing through them.

As electrochemical cells, fuel cells are comprised of two electrodes and an electrolyte, which enables ion exchange between them. The anodic and cathodic regions are separated by a polymeric membrane. The types of electrolyte are diverse and determine the performance of the cell; for example, the pressure of the hydrogen produced and the operating temperature. Electrolytes can be liquid or solid. Conventional electrolyzers use liquid alkaline electrolytes, while modern ones use solid electrolytes. The latter type is known as the proton exchange membrane fuel cell (PEMFC). Depending on the sizing of the system, different types of fuel cells can be used. For instance, the PEMFC, which is the most used technology (it operates at 80 °C), is preferable for industrial applications, as stacks of 100 kW of rated power can easily be found on the market. For high-power applications (in the range of MW), the so-called solid oxide fuel cell (SOFC) – which operates at 650 °C – is a good option, as 2 MW stacks can be found on the market.

For illustrative purposes, the electrochemical reactions in a fuel cell using PEM are depicted as follows:

$$H_2 \Leftrightarrow 2H^+ + 2e^- \qquad \text{(anode)},$$

$$\frac{1}{2}O_2 + 2H^+ + 2e^- \Leftrightarrow H_2O \qquad \text{(cathode)},$$

$$H_2 + \frac{1}{2}O_2 \Leftrightarrow H_2O \qquad \text{(total)}.$$

As can be noted, the hydrogen gas (the fuel) stored in the tanks, or in another storage medium, is reduced in the anode. The electrons delivered flow through the external load to which the system is connected, while the hydrogen protons travel through the polymeric electrolyte to the cathode. There, they combine with oxygen gas to form water. As an exothermic reaction, this also delivers heat. The thermodynamic voltage in the cell is 1.227 V. However, in practice, such a voltage is not achieved because of the efficiency of the cell. According to Harrison *et al.* [120], a PEMFC operating at standard operating conditions – that is, using hydrogen and oxygen at 25 °C and 1 atmosphere – provides a cell voltage of around 0.8 V, yielding a voltage efficiency of around $0.8/1.227 = 0.65$, or 65%.

It is important to note that there are several ways to store hydrogen [121]: it can be stored in gaseous or liquid form. The storage of hydrogen gas in metal tanks is currently the most mature, cheapest, and most reliable method. In this way, hydrogen can be stored for several hours (up to 30 h) without noticeable losses.

Modular designs of RFC can be built up to 100 MWh/10 MW storage systems. As flow batteries, the energy capacity of the system depends on the stored volume of, in this case, hydrogen. This means that RFCs are able to inject or absorb power continuously for several hours. The remarkable characteristics of this technology are its high ramp power rates, even at partial load, and its great cyclability, which is greater than the cyclability of flow batteries and conventional batteries. On the other hand, the high flammability of hydrogen gas must be properly addressed by adequate safety measures. The major drawback of the technology, however, is its low energy efficiency. Assuming energy efficiencies for the electrolyzer and the fuel cell of about 60 and 70%, respectively, the round-trip efficiency of the system falls to 42% [122].

For an overview of the various energy storage systems (ESSs) described in this chapter, see Figures 4.12 and 4.13 and Tables 4.1 and 4.2.

4.2.5   The Flywheel Energy Storage System (FESS)

Flywheels store kinetic energy in a rotating disk that is mechanically coupled to the shaft of an electrical machine. When the machine accelerates – that is, operates as a motor – energy is transferred to the flywheel and stored in the form of kinetic energy. In opposite terms, the flywheel is discharged when the electrical machine regenerates through the drive; that is, when the speed of the system is reduced [123]. Thus, the energy stored by flywheels can be expressed by

$$E_{fw}(\text{joules}) = \frac{1}{2}J\omega^2, \tag{4.6}$$

where J (in kgm^2) is the inertia of the rotating parts – that is, the flywheel itself and the rotor of the machine to which it is connected – and ω is the rotational speed, in rad/s. Figure 4.14 depicts the topology of the system.

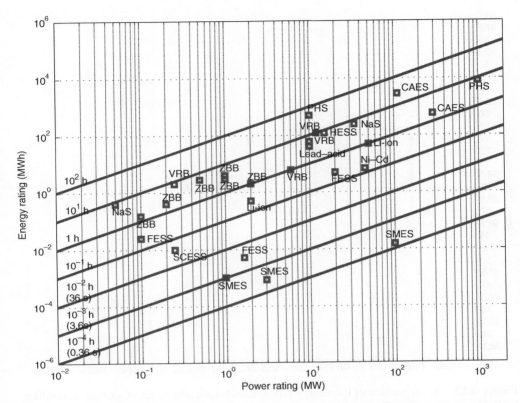

Figure 4.12 The discharge time at rated power of the considered ESSs. *Source:* Díaz-González *et al.*, 2012 [42]. Reproduced with permission of Elsevier.

The energy capacity of the system is thus limited by the maximum and minimum operating speeds of the flywheel. The power capacity is limited by the maximum torque produced at the shaft of the electrical machine, which is directly translated into an electric current. Therefore, the ratings of the electrical equipment bound the maximum or peak power of the system.

Nowadays, flywheels are high-tech systems. All rotating parts are supported by advanced magnetic bearings in order to reduce friction at high speeds. Also, with the aim of reducing wind shear, the structure is placed in a vacuum. Moreover, advanced lightweight but high-strength composite materials are used in the rotating disk. In addition, an advanced high-speed electrical machine is included in the system. Commercially, axial-flux and radial-flux permanent magnet machines are most often used in flywheels. And, finally, the electric power exchanged with the grid at the connection point of the system is run through controlled electronic power converters. All this technology is used to configure two types of flywheels, depending on the

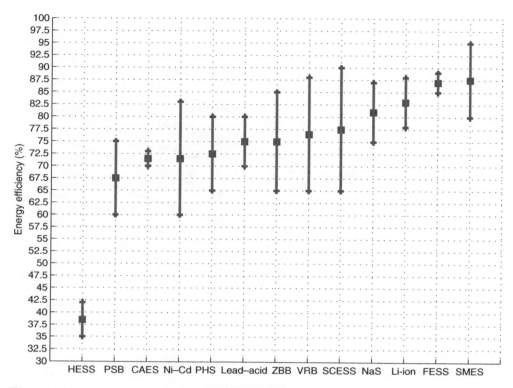

Figure 4.13   A comparison of the energy efficiencies for various kinds of storage, according to the data presented in Table 4.2.

rotational speed range. Low-speed flywheels operate in the range of thousands of revolutions per minute (rpm), while high-speed flywheels can reach speeds in the range of tens of thousands.

The major advantages of flywheels are their high efficiency (around 90% at rated power), their very long cycling life (up to 10^7 cycles), their very high ramp power rates, and their high power and energy density [115, 123]. On the other hand, the use of flywheels is limited to short-term storage applications, as the self-discharge rate of the system is around 20% of the stored capacity per hour. In fact, flywheels are only able to inject or absorb power at full load for a few minutes.

4.2.6   Superconducting Magnetic Energy Storage (SMES)

In this type of system, the energy is stored in a magnetic field. This magnetic field is created by a DC current flowing through a superconducting coil at cryogenic

Table 4.1 The specific energy and power, as well as the cycling capability and life in years, of ESSs

Technology	Specific energy (Wh/kg)	Specific power (W/kg)	Cycling capability	Life (yr)
PHS	–	–	2×10^4–5×10^4 [124]	30–50 [93], 50 [102]
HESS	100–150 [117], 400–1000 [117]	–	2×10^4 [117]	15 [117]
CAES	3.2–5.5 [107]	–	10^4–3×10^4 [124]	30 [125], 40 [126]
VRB	20 [126], 25–35 [112]	166 [127]	1000 [110], 13 000 [128]	10 [118], 15–20 [129], 20 [126, 112]
ZBB	60 [110], 70–90 [117], 75–85 [109]	45 [130]	2000 [131], 2500 [128]	8–10 [129]
PSB	–	–	–	15 [112]
NaS	100 [110], 175 [132]	115 [130], 90–230 [114]	2500 [101]	12–20 [102], 10–15 [93]
Lead–acid	30 [107], 35–50 [133]	180 [107], 200 [130]	200–300 [134], 500 [135], 1200–1800 [107], 1800 [136]	5–15 [107]
Ni–Cd	30–40 [137], 50 [107, 110], 45–80 [133]	100–150 [137], 160 [130]	3500 [138][98]	13–16 [129], 20 [133, 98]
Li-ion	80–150 [107], 100–150 [110], 160 [133], 120–200 [139]	245–430 [137], 400–500 [132], 500–2000 [107]	1500 [107], 3500 [140]	14–16 [129]
SMES	10–75 [141]	–	10^4–10^5 [223]	20 [130]
FESS	20 [117], 5–80 [142], 5–100 [107]	11 900 [143]	10^5–10^7 [107]	20 [169]
SCESS	2–5 [139], 5.69 [144], 1–10 [145], 10 [142], 5–15 [90], 30 [146]	800–2000 [90], 2000–5000 [139], 10 000 [107, 145], 13 800 [144], 23 600 [147]	5×10^5 [107], 10^6 [145]	8–10 [90], 12 [107], 17 [148]

temperatures. Superconductor materials present almost negligible resistance while at cryogenic temperatures, so the magnetic field in the coil can be created and maintained with a very small amount of current flowing through it; very little energy is dissipated by ohmic losses. Figure 4.15 presents the topology of the system.

Table 4.2 The energy efficiency, daily self-discharge, and manufacturers of ESSs

Technology	Energy efficiency (%)	Daily self-discharge (%)	Manufacturers
PHS	65–75 [93], 67 [102], 75–80 [107][126]	No [102]	First Hydro Company [56], MWH [149]
HESS	35 [90], 40 [125], 35–40 [150], 42 [122]	No [151]	Fuel Cell Energy, Inc. [151]
CAES	70 [90], 71 [126], 73 [107]	No [102]	Dresser–Rand [152]
VRB	65–75 [118], 76 [126], 75–85 [109], 78 [128], 72–88 [136]	Very low [118]	Prudent Energy Corporation [118], Vionx Energy [153]
ZBB	68 [128], 70 [154], 80 [112], 75–85 [109]	No [113]	Redflow [113], ZBB Energy Corporation [155]
PSB	60–65 [129], 75 [112]	No [112]	–
NaS	75–85 [93], 80 [98], 85 [102], 84–87 [107]	No [101]	NGK [156]
Lead–acid	75–80 [93], 70–80 [136]	<0.1 [107], 0.1 [89, 157], 0.2 [154]	Alcad [158], Exide Technologies [159]
Ni–Cd	72 [160]	0.2 [157], 0.3 [107]	Saft [161], Alcad [158], Harding Energy, Inc. [137]
Li-ion	78 [140], 88 [162]	1 [89], 5 [107]	A123 Systems [163], Li-Tec Battery GmbH. [164], Harding Energy, Inc. [137]
SMES	80 [89], 90 [223], 95 [165]	10–15 [166]	Superconductor Technologies, Inc. [167] SuperPower, Inc. [168]
FESS	85 [90]	100 [107]	Beacon Power [169], Active Power [170], Piller [171]
SCESS	65 [154], 80 [89, 286], 90 [172, 173]	5 [90], 10–20 [89]	Maxwell [148], EPCOS [174], NEC–Tokin [175]

The energy stored is determined by the self-inductance of the coil L (in henries) and the square of the electric current I (in amperes). Thus,

$$E_{\text{SMES}}(\text{joules}) = \frac{1}{2}LI^2. \tag{4.7}$$

There are two types of SMES systems, depending on the working temperature of the coil: SMES systems based on high-temperature coils (HTS) and low-temperature

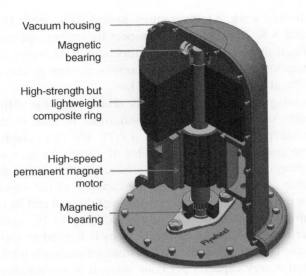

Vacuum housing

Magnetic bearing

High-strength but lightweight composite ring

High-speed permanent magnet motor

Magnetic bearing

Flywheel

Figure 4.14 The illustrative topology of a flywheel-based ESS.

Figure 4.15 The illustrative topology of an SMES system. The liquid helium is contained in the two tanks on the left, while the tank on the right contains the superconducting coil. *Source:* Nielsen, 2010 [176]. Reproduced with permission of K.E. Nielsen.

coils (LTS). The former work at temperatures around 70 K, while the latter work at temperatures around 5 K. Therefore, a key aspect for proper operation of the latter system is to maintain these low operating temperatures. However, due to the very low energy consumption of the system's cryocoolers, the energy efficiency of SMES systems is very high, at around 90% [223, 273].

The major advantage of SMES systems is related to their ability to inject or absorb vast amounts of energy in a very short time. In fact, in this regard they compete with supercapacitors (see Section 4.2.7). We can establish an analogy between the charge/discharge temporal profiles of supercapacitors and SMES systems: while the charging time constant of supercapacitors is proportional to the equivalent resistance of the electrolyte and the capacity of the supercapacitor cell, in SMES devices it is proportional to the resistance of the coil and its self-inductance.

Also, the cyclability of the system is very high, at up to 10^5 cycles at 100% of DoD. On the other hand, the use of SMES devices is limited to short-time storage applications, as the self-discharge rates of the system are relatively high, in the range of 10–15% of the rated energy capacity per hour. Also, it is worth noting that this type of storage device becomes completely discharged in a very short time, discharging at full load (in the range of seconds up to few minutes).

Intensive R&D activities are still pending around SMES, with the objective of making this technology technically and economically implementable. Few demonstration projects can currently be found worldwide. One example is that carried out by SuperPower, Inc., in collaboration with ABB, Inc., Brookhaven National Laboratory, and the Texas Center for Superconductivity, at the University of Houston [168]. The project was focused on the development of a 20 kW/MJ SMES device.

4.2.7   The Supercapacitor Energy Storage System

Supercapacitors are based on electrochemical cells that contain two conductor electrodes, an electrolyte and a porous membrane that permits the transit of ions between the two electrodes. Thus, the presented layout is similar to the electrochemical cells of batteries. The main difference between supercapacitors (or ultracapacitors, or double-layer capacitors, depending on the literature) and batteries lies in the fact that no chemical reactions occur in the cells, but the energy is stored electrostatically in the cell [145, 177].

In supercapacitors, the electrodes and the electrolyte are electrically charged (the cathode is positively charged, the anode is negatively charged, and the electrolyte contains both positive and negative ions). At each of the electrode surfaces there is an area that interfaces with the electrolyte, and it is in each of these areas where the phenomenon of the "electrical double layer" occurs.

By applying a voltage between the electrodes, both the electrodes and the electrolyte become polarized. This means that the positive charge of the cathode is transferred

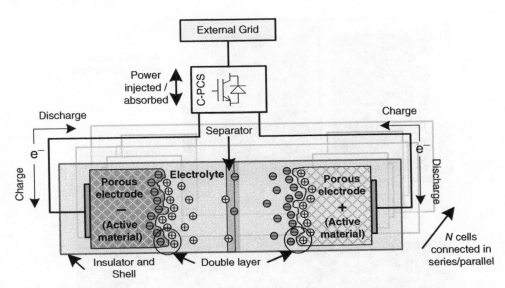

Figure 4.16 The illustrative topology of a supercapacitor, depicting the electrical double layers at each electrode/electrolyte interface.

to the area interfacing with the electrolyte, forming a layer of positive ions. In turn, the negative ions of the electrolyte are transferred to the same electrolyte/cathode interface, forming a negative charge-balancing layer of ions. These two layers build up an "electrical double layer." The mechanism behind the operating principle of such a double layer can be explained using the Helmholtz model. The model establishes that the two layers are separated by a layer of solvent molecules of the electrolyte, called the inner Helmholtz plane. This layer of solvent molecules actually separates the positive and negative charges of the electrode and electrolyte, thus acting as a dielectric. Ultimately, there is a potential difference between the two layers of positive and negative ions derived from the electric field within them, and the double layer can be taken to resemble a capacitor (the described double-layer concept can be observed in Figure 4.16; see also Figure 4.17).

Therefore, the magnitude of the electrical potential V (in volts) between the two layers of positive and negative ions at each electrode/electrolyte interface, in conjunction with the resultant capacitance F (in farads), determines the energy stored in the supercapacitor. Thus,

$$E_{sc}(\text{joules}) = \frac{1}{2}CV^2. \tag{4.8}$$

The voltage generated in the cell is dependent on the strength of the electric field between the layers building up each of the "electrical double layers" described above.

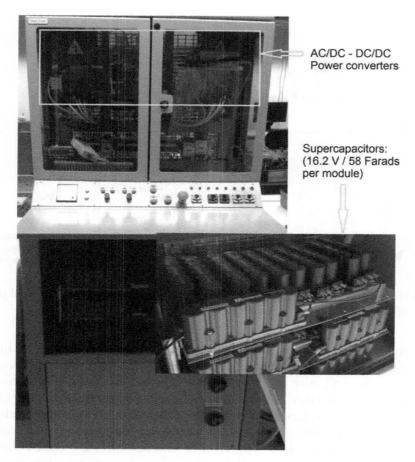

AC/DC - DC/DC
Power converters

Supercapacitors:
(16.2 V / 58 Farads
per module)

Figure 4.17 Supercapacitor modules from Maxwell Technologies, Inc., in a configuration on a test bench in the IREC laboratory. *Source:* Adapted from IREC, 2015 [108].

This electric field is, in turn, proportional to the amounts of positive and negative ions located at the electrode/electrolyte interface. So to avoid transfer of ions between the two layers of positive and negative ions, thus decreasing the voltage within the double layers, the breakdown voltage of the dielectric should be maximized. As noted before, this dielectric is provided by solvent molecules of the electrolyte. In this way, the selection of the electrolyte is key to ensuring the maximum energy capacity. Usually, both aqueous and organic electrolytes are commonly found, the latter being the most common type. With aqueous electrolytes, a cell voltage of around 1 V can be obtained, while it can be increased up to 2.5 V by using organic types [178].

As stated in Equation (4.8), the second factor affecting the energy capacity of supercapacitors is the capacitance of the cell. The capacitance C (in farads) of a capacitor is given by the quotient between the stored charge Q (in coulombs) per unit of voltage V (in volts), so

$$C = Q/V. \tag{4.9}$$

In addition, it can be expressed as a function of the permeability of the dielectric, its thickness, and the area holding each of the layers of the electrical double layer. Then,

$$C = \varepsilon\varepsilon_0\frac{A}{d}, \tag{4.10}$$

where ε is the dielectric constant, ε is the permittivity of a vacuum, A is the effective area of the surface of the electrode, and d is the dielectric thickness.

In order to maximize the capacitance, different metal-oxide electrodes, electronically conducting polymer electrodes and activated carbon electrodes, are used in industry. These materials are porous, so they can maximize the effective area of the electrode in which ions can be allocated. The most common types are the ones based on activated carbon, since they can lead to supercapacitors with a high energy density and capacitances around 5000 F [145]; that is, capacities up to 1000 times per unit volume more than those of conventional electrolytic capacitors.

With regard to the distribution of capacitance between the two electrical double layers in the cell, we can distinguish between symmetrical and unsymmetrical supercapacitors. Symmetrical ones are those with the same effective area in both electrodes. Since the cell can be considered to resemble two capacitors in series (given by the two double layers at each electrolyte/electrode interface), the total capacitance can be formulated as

$$C_{\text{eq}} = \frac{C_1 C_2}{C_1 + C_2}, \tag{4.11}$$

where C_1 and C_2 are the equivalent capacitances in each electrical double layer.

As mentioned, the electrolyte and electrode materials have a fundamental influence on the energy and power capacity of the supercapacitor, and also on its dynamic behavior. To be precise, and with reference to the supercapacitor dynamics, one defining parameter is the so-called charge/discharge time constant, τ. This is given by the product of the equivalent series resistance (ESR) of the supercapacitor and its capacitance. Thus,

$$\tau = RC. \tag{4.12}$$

The time constant is the time needed to discharge 63.2% of full capacity with a current limited only by the internal resistance – or the ESR, as it is commonly known – of the

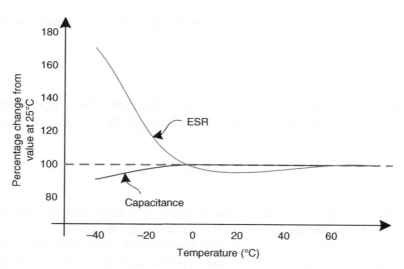

Figure 4.18   The capacitance and the ESR as temperature-dependent characteristics. *Source:*
Adapted from Maxwell Technologies, Inc., 2015 [148].

supercapacitor. The ESR weights the losses in the supercapacitor while charging and
discharging; that is, those associated with the movement of ions within the electrolyte
and across the separator. The ESR is normally in the range of milliohms [179] and is
a temperature-dependent parameter, as presented in Figure 4.18.

Apart from the ESR and the capacitance, the third characteristic parameter for the
supercapacitor is the leakage resistance, which weights the self-discharge of the cell.
This resistance is much higher than the ESR. All three parameters – the capacitance,
the ESR, and the leakage resistance – can be found in manufacturers' datasheets,
and from them, averaged models for supercapacitors can be built (as presented in
Chapter 6.

In summary, supercapacitors are characterized by offering high ramp power rates,
high cyclability (comparable with the cyclability of flywheels), high round-trip efficiency (of up to 80%), and a high specific power, in W/kg, and power density, in W/m^3
(10 times more than for conventional batteries). The latter characteristic defines supercapacitors as well suited for applications that impose major volumetric restrictions.

On the other hand, major drawbacks of the technology are related to its high self-
discharge rates (of up to 20% of the rated capacity in only 12 h) and its limited
applicability to situations where high power and energy are needed. In fact, the development of supercapacitors is mostly focused in fields such as automotive and portable
devices. Finally, it is worth noting that as a short-timescale ESS, supercapacitors are
unsuitable in that they are expensive in comparison with other competitors such as
flywheels. Their cost is estimated as 10 times the cost per kWh of flywheels.

Supercapacitors are, in general, young technologies. The first prototypes were developed in 1957 by H.I. Becker (General Electric). However, the first related studies were carried out in the nineteenth century by Helmholtz, who discussed the electrical behavior of a metal surface while immersed in an electrolyte. Currently, intense research activity is under way to scale up supercapacitor size, and to improve their performance, so that they will be suitable for both stationary and nonstationary applications – such as, for instance, in the field of electromobility.

4.2.8   Notes on Other Energy Storage Systems

To extend the catalog of storage technologies previously presented, this section tangentially approaches the field of thermal storage. In particular, the case of thermal storage in molten salts, widely used in solar plants and in the field of renewable generation, is explained. The second technology to be approached here is the so-called "power-to-gas" (P2G) concept. Again, its description is included here since – apart from attracting much attention from industry and the energy sector in general – this technology can be associated with the storage of wind power.

4.2.8.1   Molten Salts as a Thermal Storage Medium

Thermal energy is usually stored at the moment of being produced, so as to avoid energy losses due to conversion of other types of energy (e.g., electrical energy) to thermal energy. So by thermal energy storage technologies, we mean systems that absorb, store, and release thermal energy, in a controlled manner, for application for other purposes. The field of thermal energy storage is very extensive. Nevertheless, we can classify the available technologies into three main categories: sensible heat media, latent heat media, and chemical heat media [180].

The energy storage capacity in sensible heat media depends on the specific heat characteristic (c_e, expressed in kJ/kgK) of the particular medium – solid, liquid, or gas – which is properly and thermally isolated so as to reduce heat losses. Therefore, the regulation of the stored energy is through the management of the mass media, and also through its temperature. For example, a sensible storage medium can be as primitive as a heated rock. Imagine that it is contained inside an oven, which is turned off during the night. In these conditions, the progressive temperature decay of the rock during the night is directly translated into thermal energy that can be used to heat the food inside the oven. Sensible heat media are the most utilized thermal storage systems due to the relative simplicity and maturity of the technology [181].

While thermal energy transfer mechanisms in sensible heat media are based on temperature variations, the temperature is kept constant in latent heat media. Here, the thermal energy is released or stored by a material during phase change processes (e.g., from solid to liquid) [182]. Finally, chemical heat storage media are those based on exothermic chemical reactions in a substance that, as a result of that process, is separated into two components. The resultant two components can be associated again by reversing the process and applying heat, thus performing an endothermic reaction.

In solar power plants, molten salts are widely used as sensible heat media. There are several practical examples of such systems worldwide [181]. Normally, the salts utilized can be either organic or inorganic: examples include sodium nitrate, potassium nitrate, and lithium salts. The operating temperature of the salts, and therefore, of the thermal storage system, oscillates between 292 and 386 °C [181]. The particular example of the Andansol I to III solar power plants (in Granada, Spain) is reported in Solar Millennium AG [183]. As presented in that article, 28 500 tons of molten salts are heated, using part of the heat collected from the sun by the solar field and transmitted by a heat transfer fluid (HTF) to a HTF/molten salts heat exchanger. Through this exchanger, "cold" molten salts from one of the two storage tanks are heated and stored in the "hot" molten salts tank. This hot molten salts are used to heat the HTF, which in turn is used to run the turbine–generator system and therefore generate electricity. Figure 4.19 shows the system topology.

In this way, thermal storage can be used in the plant for various purposes. Amongst the possible applications, the stored thermal energy can be used to prevent steam turbines from being shut down due to intermittent steam production from solar isolation

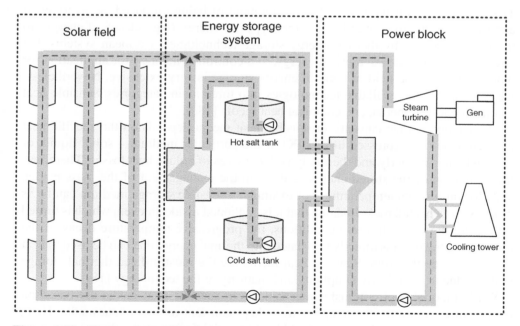

Figure 4.19 The topology of a solar power plant with a storage system based on molten salt. *Source:* Adapted from Solar Millennium AG, 2015 [183].

in cloudy periods. Also, the stored energy can be used to run a steam turbine during hours with no sun, thus time-shifting the electricity production when required. In the particular case of the Andasol plant, the capacity of the storage system permits the plant to continuously generate electricity for up to 7.5 h of no sun.

4.2.8.2 The Power-to-Gas Concept

Power-to-gas refers to the conversion of electricity (normally from renewable generation) into a gas that can be used as a fuel in various applications (e.g., the tertiary sector, the gas grid, mobility, and industry). This gas can be hydrogen or synthetic methane, which can be delivered to final users by relying on the huge gas transport capacity of gas pipelines (in the order of TWh in energy) [184]. This concept is considered one of the most promising strategies for renewable integration and decarbonization of large interconnected power systems.

When hydrogen gas is obtained, it is produced by an electrolyzer, and hence by means of water electrolysis (see Section 4.2.4). In this case, an excess of renewable generation that cannot be injected into the power system can be transformed into chemical energy, in the form of hydrogen gas, which in turn can be stored in tanks to be used later for other purposes (e.g., as fuel for a fuel cell to produce electricity again, or as a fuel gas to fire a gas turbine). Moreover, the hydrogen can be used not as a gas for generating electricity, but as an industrial fuel and for vehicles, for example, or it can be injected into natural gas pipelines for transport for other uses. However, this last option is limited by the fact that due to the high flammability of hydrogen, its total volume stored in pipelines is limited by European regulations to between 0.2 and 12% of the total natural gas transport capacity, depending on the country [185].

Moreover, hydrogen can be used, in conjunction with CO_2, to produce synthetic methane by means of a catalytic process called methanation. This is based on the Sabatier principle, and yields the following chemical reaction [184]:

$$CO_{2(gas)} + 4H_{2(gas)} \Leftrightarrow CH_{4(gas)} + 2H_2O_{(liq.)}.$$

As can be noted, carbon dioxide and hydrogen gas react to create methane and liquid water. The process is exothermic, yielding 165 kJ/mol of thermal energy per mass unit. The efficiency of this process proves to be between 70 and 85%, with the remainder emitted as heat [186]. The carbon dioxide can be obtained from the atmosphere, from biomass or biogas, or fossil fuel driven power plants equipped with carbon capture.

The overall energy efficiency can vary dramatically depending on the number of energy conversions involved in the process. As a figure of merit, it is worth noting that the electricity/hydrogen gas conversion efficiency is around 70%;

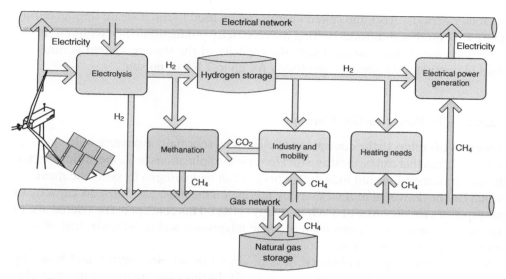

Figure 4.20   The power-to-gas concept. *Source:* Adapted from Grond, Schulze, and Holstein, 2013 [186].

electricity/electricity using hydrogen gas can fall to 40% (see Section 4.2.4); and electricity/hydrogen gas/methane gas conversion results in around 50–60%. These percentages also depend on the storage media (e.g., metallic tanks in which hydrogen can be stored at pressures up to 350 bar [42]), on the losses incurred during transport in gas pipelines, and on the technology of electrolyzers, amongst other factors. Figure 4.20 graphically depicts the power-to-gas concept explained above.

This technology, although very attractive from the point of view of the design and operation of large power systems, is still incipient. However, several demonstration projects can be found at the European level. The largest was commissioned in 2013 by Solar Fuel GmbH, in Germany, at the behest of vehicle manufacturer Audi AG. This plant produces hydrogen and synthetic methane from electricity, water, and carbon dioxide (which comes from a nearby biogas plant). The gas produced is to be injected into the local gas distribution grid, with the aim of feeding compressed natural gas for vehicles. The capacity of the installation can reach up to 6.3 MW in electricity consumption, yielding 360 Nm^3 per hour of synthetic methane [186].

As further proof of the interest that this technology is attracting, we note here European platforms such as dena's Strategieplattform Power to Gas [187] (in Germany) and the North Sea Power to Gas Platform [188], which aim to conduct dialogs with politicians, disseminate the concept of the technology, and promote collaboration between industry, associations, and science, altogether around the power-to-gas concept.

4.3 Power Conversion Systems for Electrical Storage

The power electronics used are diverse and depend on the type and final application of the storage system. Accordingly, the objectives of this section are twofold: (i) to provide a general view of the topology of power conversion systems, addressing the type of storage container; and (ii) to present actual examples of representative power conversion systems utilized in different fields of application. Finally, some notes on battery management systems (BMSs) are presented, since they are closely related to the power electronics discussed here for batteries, as they act as an interface between the storage technologies and the electrical systems to which they are connected.

4.3.1 Application: Electric Power Systems

4.3.1.1 A General Description

Those storage technologies not synchronized with the network are connected through and managed by power electronics. Regardless of the specificities of the various storage technologies, the power conversion systems for all forms of storage can be made to resemble the topology presented in Figure 4.21. As can be noted, the power conversion systems for all kinds of storage include a common part, composed of switchgear, a coupling transformer, an inductive filter, and the so-called grid-side converter (GSC). This converter interfaces the AC grid voltages with the DC link. The aim of this converter is to ensure a constant and stable voltage at the DC link, to ensure that the control algorithm of the storage-side converter (SSC) works properly.

The topology of the SSC can vary between a three-phase inverter and a buck DC–DC converter. The former relates to flywheels, while the latter is employed for

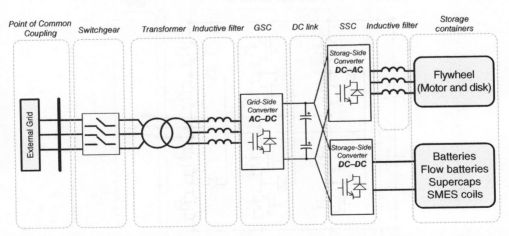

Figure 4.21 The normal topology for power conversion systems for storage not synchronized with the network.

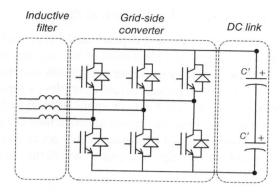

Figure 4.22   The normal topology for the GSC.

batteries, flow batteries, supercapacitors, and SMES devices. The aim of the SSC is to actually manage the electric power absorbed or delivered by the storage container.

The type of the GSC usually corresponds to a three-phase, two-level H-bridge structure, as depicted in Figure 4.22. This mature architecture is widely proven for configuring bidirectional power converters that can handle a few megawatts, assuming that both the AC and the DC voltage levels are in the range of kV. To further increase the ratings of the converter, such a structure should consider paralleling and/or serializing power switches (which are usually IGBTs).

There are other options, though, such as multilevel converters [189, 190] (see, e.g., Figure 4.23). The main advantage of these converters compared to two-level

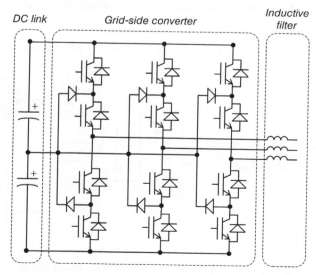

Figure 4.23   The three-level neutral point clamped inverter.

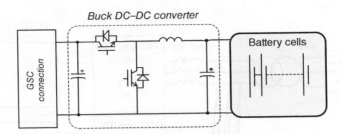

Figure 4.24 Buck DC–DC converters for energy storage.

H-bridge structures is the ability to provide more output voltage levels, higher voltage amplitudes, and a larger output power.

Furthermore, another interesting option is those based on multiple module converters. The idea behind this concept is to connect several modules or converters in parallel and/or in series to increase their capability and reliability. The cascaded configuration is also known as the modular multilevel converter (MMC). It consists of several interconnected modules, each one of which, configured into a single-phase H-bridge (or half-bridge), has its own capacitors, enabling different voltage levels on each module. It is worth remarking that such designs are well established on the HVDC transmission systems market, as they allow us to reach tens and even hundreds of kV [191]. ABB, Siemens, and Alstom have their own concepts, called HVDC Light, HVDC plus, and HVDC MaxSine, respectively. Their application in the field of storage is still under discussion.

As previously noted, the type of SSC varies between DC–AC and DC–DC structures depending on the storage container. For flywheels, the SSC can present the same structure as for the GSC. In this case, both the GSC and the SSC would be used to configure a symmetrical AC–DC–AC power conversion system, with two converters in a back-to-back arrangement.

Conversely, for DC–DC SSCs, the type of converter can vary depending on the final application. Accordingly, some structures are described in the following sections. Buck DC–DC architectures are widely employed for storage in electric power systems (see Figure 4.24).

4.3.1.2 Examples of Real Systems

Normally, the energy and power capacity needed for storage systems in applications related to power systems are huge. For instance, the storage facility would be required to exchange tens of MW over a period of several hours while being supposed to compensate mismatches between generation and demand in the network.

Such huge outputs can be provided by a few particular forms of storage, such as pumped hydroelectric installations. However, there is no battery or flywheel on

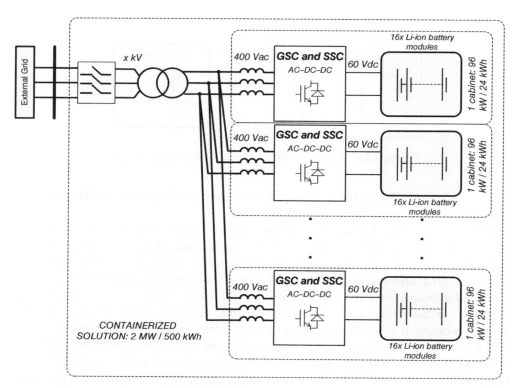

Figure 4.25 A graphical representation of the Siemens SIESTORAGE containerized solution. *Source:* Adapted from Siemens AG, 2015 [192].

the market able to match this performance. Thus, modular and scalable designs are necessary to reach the required ratings, and these are just realizable with the proper power conversion system design.

On the market, we can find several modular battery-based solutions, designed to be connected in medium-voltage networks. Siemens' example [192] is called the SIESTORAGE system. This system packs up to 16 Li-ion battery modules, reaching up to 96 kW/24 kWh, into a cabinet. The output voltage of the battery modules is around 60 V DC, so the cabinet includes an inverter, which steps up the voltage to 400/230 V AC. Then, a coupling transformer can be connected to further increase the voltage to the MV range in order to connect the system to the power system. A schematic of the system is shown in Figure 4.25.

The rating of the system can be further increased by paralleling the cabinets described above. The resultant is a containerized solution rated at up to 2 MW/500 kWh of power and energy capacity.

Such modularity and scalability can be also observed in Alstom's example [193], called MaxSine™ eStorage. This system can reach 1.25 MW per module. A project reference using this technology is reported by the manufacturer [194]. In this case, the MaxSine™ technology was used to configure a 1 MW/560 kWh installation at the primary substation of a microgrid in the municipality of Carros (French Riviera), for peak-load management.

ABB's alternative is called DynaPeaQ® [195]. This system is based on voltage source converters building flexible AC transmission systems (FACTS) for application in transmission and distribution networks. ABB claims that ESSs based on this technology can be rated at 50 MW/50 MWh and beyond. These ratings are not achieved by paralleling containerized independent modules rated at 1–2 MW, all equipped with storage and dedicated power inverters (as in the Siemens and Alstom examples). Instead, a single power converter can be rated at a multi-megawatt scale by proposing the use of converter topologies employing series-connected high-power IGBTs. Doing this, power converter branches can withstand higher electrical potentials between their terminals, thus increasing the power capacity of the device. Such multi-kilovolt electrical potential on the DC side of the GSC is supposed to be directly provided by strings of several Li-ion batteries connected in series, so SSCs are not needed.

An example of the application of this system is reported in Wade *et al.* [196]. This work, written prior to the commissioning of the system, discussed the optimal location and operation. The storage system was rated at 200 kW/200 kWh (so in this case, the authors were not configuring a multi-megawatt system). The ratings of the converter were increased to 600 kVA to also exchange reactive power. The final application of the system is in a meshed medium-voltage network (11 kV) with distributed generation. The storage system is composed of a string of Li-ion battery modules, which provide up to 5.8 kV DC. An inverter and a step-up transformer connect the system to the electrical network. Figure 4.26 depicts the topology of the system.

To complete the description of actual examples of power conversion systems for storage in the electric power system, the following paragraphs present a modularized

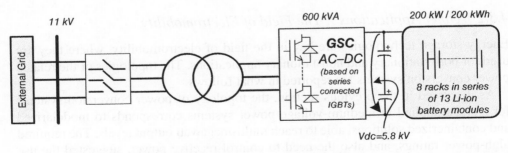

Figure 4.26 ABB's example of a multi-megawatt medium-voltage energy storage solution. *Source:* Adapted from Wade *et al.*, 2009 [196].

Figure 4.27 A flywheel-based storage plant for frequency control (20 MW/5 MWh). *Source:* Beacon Power, LLC, 2015 [169]. Reproduced with permission of Beacon Power, LLC.

solution for flywheel-based storage installations. The adopted example is a flywheel plant for frequency regulation in New York, commissioned by Beacon Power [169]. The system is comprised of 200 flywheel units of 100 kW/0.025 MWh, connected in parallel to increase the ratings of the plant up to 20 MW/5 MWh. Figure 4.27 shows the actual installation.

As mentioned above, all 200 flywheel units are connected in parallel. Accordingly, the power conversion system of the storage plant is comprised of 200 AC–DC–AC bidirectional power converters, one per flywheel unit, all connected on the grid side in a single common coupling with the external network.

4.3.2   Other Applications I: The Field of Electromobility

Energy storage technologies are key in the field of electromobility, where they are used for both stationary and nonstationary applications. The topologies of the related power conversion systems are reported in what follows.

As discussed in the previous section, the topology of power conversion systems for storage applied in medium-voltage power systems corresponds to modularized and containerized solutions, able to reach multi-megawatt output levels. The required high-power ratings, and also the need to control reactive power, suggested the use of high-power IGBTs for power converters. Moreover, all power conversion systems presented so far are bidirectional, so as to charge and discharge the storage system as requested by the network.

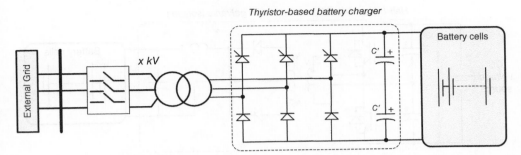

Figure 4.28 A thyristor-based battery charger and DC power supply system. *Source:* Adapted from Schaefer, Inc. catalog; Schaefer, Inc., 2015 [197].

Now, looking at power converters for storage in the field of electromobility, battery chargers based on thyristors, which are able to exchange hundreds of kilowatts, are also usually found. Unidirectional power conversion systems for battery chargers can also be considered. For instance, such battery chargers are applied at both railway onboard and trackside sites. An example of such a system is reported by Schaefer, Inc. [197]. Briefly, and as figures of merit, it is worth noting that the input voltage of such a converter is about 230 V (single-phase) / 480 V (three-phase) AC, depending on the system. On the DC side, the output voltage varies from 12 V DC up to 400 V DC. The output (DC-side) current can reach 3250 A, altogether bringing the power up to 500 kW. The topology of the converter is shown in Figure 4.28.

One important remark at this point is that the blocking of thyristors is performed when the waveform of the current flowing through the device is zero, so it is not fully controlled as in the case of IGBTs. Therefore, ultimately, the alternating currents on the AC side of the converter may not be in phase with the voltage waveform, resulting in reactive currents. In this way, both active power factor correction (switching algorithms) and passive strategies (the inclusion of ancillary components such as shunt capacitor filters) need to be applied for this purpose.

In the portfolio of the manufacturer SMA [198], battery chargers for the field of electromobility can also be found [199]. These chargers interface the battery with the DC power supply system of the train or catenary. Therefore, this DC–DC power converter steps the voltage input (e.g., around 530 V DC) down to suitable levels for battery charging (e.g., 90 V DC). Since protection is needed against short circuits, this DC–DC converter is based on a hard-switching half-bridge topology with galvanic isolation, provided by a medium-frequency transformer. Figure 4.29 depicts such a topology.

Apart from trains, one increasingly important market nowadays for battery chargers in the field of electromobility is electric vehicles. For this purpose, there are several types of battery charger, depending on the charge applied (fast or slow, one-directional

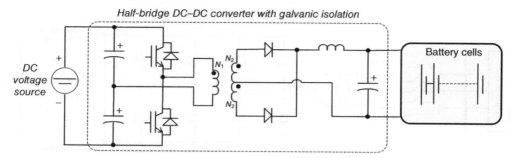

Figure 4.29 An IGBT-based half-bridge DC–DC converter with galvanic isolation.

or bidirectional) and the type of vehicle (hybrid, pure electric, fuel cell, etc.), amongs other factors.

Battery chargers for electric vehicles can be unidirectional or bidirectional. With regard to unidirectional topologies, the simplest version can be comprised of a step-down low-frequency transformer, coupling the main low-voltage grid with a single-phase (or three-phase) full-wave diode rectifier, equipped on the DC side with a step-down chopper [200]. Although cheap, such a converter does not provide controlled battery charging, thus yielding poor performance and also reducing the battery life. Figure 4.30 shows the topology of the converter.

While in the charging station, an electric vehicle can be viewed by the network operator as a storage capability, so the operator may make use of the energy stored in the battery to provide services to the grid. If so, bidirectional charging points are needed, enabling bidirectional power flows between the grid and the vehicle. All these related aspects are known as "vehicle-to-grid." Suitable topologies for bidirectional chargers can be those presented in Figures 4.31 and 4.32.

As can be noted, in Figure 4.31, the converter is coupled to the grid through a LV/LV low-frequency transformer, which provides galvanic isolation for device protection against short circuits in the network. The AC low voltage is rectified in a three-phase

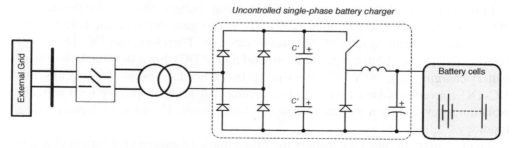

Figure 4.30 An uncontrolled single-phase unidirectional battery charger. *Source:* Adapted from Davis *et al.*, 1999 [200].

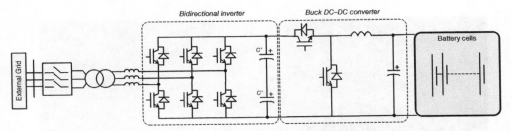

Figure 4.31 A vehicle-to-grid charging point topology with a buck/boost DC–DC converter.

IGBT two-level inverter, which in turn is coupled to a buck/boost DC–DC structure to step the voltage down to the required levels for battery charging and discharging. The use of IGBTs enables the management of bidirectional power flows in the above-mentioned scheme. One option to avoid the use of the buck/boost converter on the DC side of the arrangement is to step down the AC voltage prior to feeding the AC–DC inverter, as depicted in Figure 4.32. However, this bounds the magnitudes of the AC voltages that can be obtained on the AC side of the inverter. In this case, a transformer is used to step the voltage up to suitable levels for integration into the grid. The values presented in Figure 4.32 correspond to a laboratory-scale test bench in the IREC laboratory (see Figure 4.33).

4.3.3 Other Applications II: Buildings

Electrical energy storage also finds several applications in buildings, such as in configuring uninterruptible power supply systems (UPSs) for critical loads. The importance of such systems is intensified, for instance, in data centers. Such installations are characterized by having to manage a huge amount of consumption from critical computers (e.g., web servers, bank databases, etc.), that have to be fed and cooled 24/7. In this

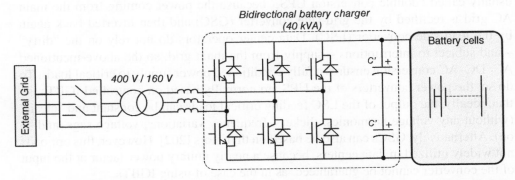

Figure 4.32 A vehicle-to-grid charging point topology without a buck/boost DC–DC converter.

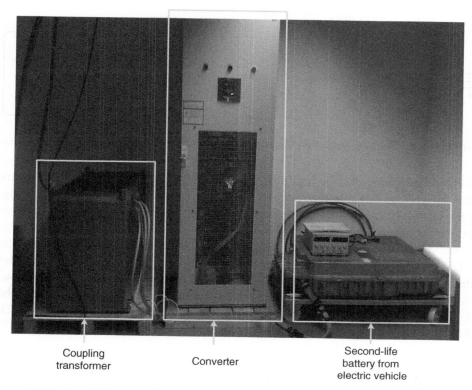

Coupling
transformer   Converter

Second-life
battery from
electric vehicle

Figure 4.33 A vehicle-to-grid charging point topology without a buck/boost DC–DC converter: the IREC test bench for research into second-life batteries from electric vehicles.

way, UPSs are key for ensuring the required performance and service level of the installation [201].

An example of the topology of UPSs is presented in Figure 4.34. This topology is usually called "double conversion UPS," because the power coming from the main AC grid is rectified by the grid-side converter (GSC) and then inverted back again by the load-side converter (LSC). Data center operators do not rely on the "dirty" – and subject to interruptions – supply from the main grid, so the above-mentioned AC–DC–AC conversion ensures a fully conditioned power supply to critical loads. To do so, the power converters of the UPS are normally based on controlled IGBTs, so that, ideally, the output of the LSC feeding critical loads can be constant and "clean" (without any voltage harmonics, flickers, frequency variations, voltage sags, and so on). Alternatively, GSCs can also be based on thyristors [202]. However, this option is not widely utilized in data centers, because a nearly unitary power factor at the input of the converter cannot be guaranteed, as in the case of using IGBTs.

In normal operating conditions – that is, not experiencing mains failures – the energy stored in the batteries should not be utilized. Therefore, they remain steadily

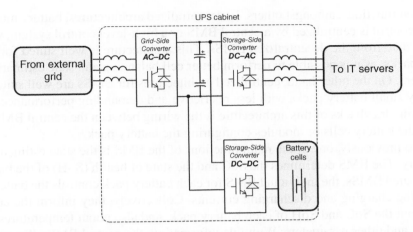

Figure 4.34   The topology of a double conversion uninterruptible power supply (UPS) system.

charged and there is no power flow through either the batteries or the DC–DC power converter interfacing the system with the DC link between the GSC and the LSC. However, in the event of a mains failure, the energy stored in the batteries is fully discharged through the DC–DC and LSC, so that the critical loads are not disrupted.

Finally, it is important to note that the DC–DC converter may be avoided, provided that the voltage of the battery bank fits with the voltage requirements at the DC link. If not, the DC–DC should be included, acting as a bidirectional buck/boost converter.

The ratings of double conversion UPSs can vary between tens of kVA up to hundreds of kVA [202] and as building equipment, they are rated at low voltage (230 V AC single-phase/400 V AC three-phase).

4.3.4   The Battery Management System (BMS)

The BMS is the system in charge of carrying out several functions related to battery data acquisition, state monitoring, and control, so as to ensure proper and safe operation during the battery's lifetime [203]. Therefore, the BMS is equipped with digital and analog inputs and outputs to read and evaluate external signals such as voltages and temperatures, and also to govern the power electronics of the battery to perform charging and discharging processes. In fact, BMSs control parameters such as charge and discharge parameters (mode, current, end-of-charge voltage, pulse current, DoD, etc.), temperatures, and maintenance actuations, amongst others. By controlling all these aspects, the BMS has a great influence on the battery lifetime [204].

The technology for building up BMSs varies with the battery pack type, size, and final application. From a systems point of view, we can identify various architectures for BMSs within a BESS, centralized and decentralized. In decentralized architectures, each battery is equipped with a BMS performing precise data acquisition and state estimation functions, amongst others. In decentralized architectures, battery monitoring and control is centralized by a central BMS or a high-level control system, which receives data from the decentralized BMSs. This architecture is well suited for large battery packs, as it is possible to perform better cell balancing and state estimation of the system. On the other hand, centralized architectures for BMSs are well suited for relatively small battery packs with less restrictive and demanding performance. One of the main drawbacks of this architecture is the wiring between the central BMS and each of the battery cells or modules configuring the battery pack.

As said previously, one of the main functions of the BMS is the state estimation of the battery. The BMS determines the SoC and the state of health (SoH) of the battery. In distributed BMSs, the individual BMS for each battery pack controls the battery by varying the charging and discharging currents. Collectively, they inform the central BMS about the SoC and SoH of each battery pack, and also about temperatures, cell voltages, and other parameters. With this information, the central BMS allocates the power demand for each battery pack, and computes individual setpoints for distributed BMSs to improve the system performance, by minimizing temperature gradients across battery cells, protecting the system from internal degradation and capacity fade, ensuring optimal charging patterns, balancing cell charges, calculating electrolyte flow in flow batteries according to the power demand, and so on. Therefore, the central BMS is equipped with an optimizer algorithm which, apart from the already detailed information, inputs exogenous signals such as those related to the final application of the system (e.g., the active power demand for load balancing). This optimizer is based on an embedded battery model, and the optimization algorithm used can be of different nature, including predictive and adaptive types. Figure 4.35 depicts a BMS architecture as usually deployed in modular battery-based designs.

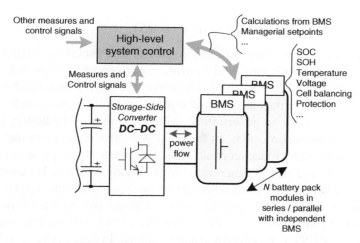

Figure 4.35   The concept of a BMS.

4.4 Conclusions

In this chapter, the operating principles as well as the main characteristics of several storage technologies and their related power conversion systems have been described. In conclusion, it is worth making the following final remarks:

- Currently, there is an extensive catalog of storage technologies covering systems of various natures and with various operating principles. The performance of all kinds of storage can be expressed in terms of principal characteristics such as power and energy capacity, cyclability, time response, and efficiency, amongst others. There are no ideal storage media – each technology is best suited for different systems and for different purposes – so any technology selection process should be based on the above-mentioned metrics.

- Power conversion systems play a key role in the development and deployment of storage technologies. The chapter has summarized the normal topologies for those types of storage that are not synchronized with the external grid, but are connected through power electronics. This review has demonstrated that power conversion systems need to be adapted to the characteristics of the storage media, and also to the final application of the system. Finally, to highlight the importance of power electronics, the brief industrial benchmark carried out has demonstrated that, indeed, scalability and modularity are currently driving the development of containerized industrial solutions reaching multi-megawatt capacities.

- Nowadays, tremendous efforts are being made to improve the capabilities and efficiencies of the available storage technologies, as well as to reduce their capital costs. The aim of this research is to make ESSs economically suitable for different fields of application.

5

Cost Models and Economic Analysis

5.1 Introduction

The techno-economic feasibility assessment of energy storage technologies, as for any other type of technology, is subjected not only to aspects inherent to the technology itself (e.g., capital costs, energy efficiency, maturity, and so on), but also to exogenous factors related to final application (e.g., electricity price, regulations and limitations for usage, etc.). Therefore, one should take all these factors into account so as to come up with the right decision on whether or not the inclusion of an storage system is beneficial for the provision of particular services in specific environments.

Ultimately, the total expenditure of the project should take into account aspects such as the capital costs of the technology, the operating and maintenance costs, and the possible replacement costs of components during the lifetime of the system, as well as those costs derived from the disposal and recycling processes. The sum of all these factors yields the so-called "total life-cycle cost" of the system. Such a total cost is preferably expressed in annualized form, to give yearly figures over the entire life span of the system. In this way, these annualized costs can be intended as the expenditure that the system operator is supposed to pay yearly, considering all of the operations around the technology (e.g., purchase, installation, operation, repayment of loans, and interest). Since the value of money will vary throughout the life span of the project, the annualized costs are levelized, and hence adjusted by taking into account future costs at a predicted discount rate. The discount rate is, precisely, the most critical parameter for determining important financial metrics of the project such as the net present value and the internal rate of return. For decision-making, and while performing financial evaluations for investment, the discount rate should be defined

Energy Storage in Power Systems, First Edition. Francisco Díaz-González, Andreas Sumper and Oriol Gomis-Bellmunt.
© 2016 John Wiley & Sons, Ltd. Published 2016 by John Wiley & Sons, Ltd.

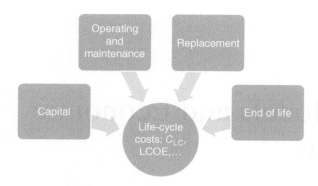

Figure 5.1   The concept of life-cycle costs.

by the investor and, in general terms, this decision should take into consideration the risk that money may be lost in investment and in terms of the opportunity cost of the project compared to other alternatives. In any event, the discount rate is based on market interest rates and can be nominal or real; that is, considering the effect of inflation rates or not, respectively. Nominal discount rates are preferably adopted by developers, while real discount rates are preferred by governments and policy-makers.

Taken together, all this makes up the "annualized life-cycle cost" C_{LC} (in €/kW-yr), as we call it hereinafter. Note that this magnitude is expressed per unit of power delivered by the storage system. Figure 5.1 graphically depicts the concepts presented above.

For decision-makers, the annualized life-cycle cost thus becomes an useful metric with which to address the economic feasibility of the project. Moreover, dividing C_{LC} by the number of yearly operating hours of the system, one can determine a second metric, which is the widely utilized "levelized cost of energy," LCOE, expressed in c€/kWh. The LCOE gives the economic resources that the storage operator (e.g., a utility), needs to charge the storage system per energy unit that it delivers, hence covering all costs around the technology. The LCOE is thus calculated as follows:

$$LCOE = 100\frac{C_{LC}}{n \times h}, \tag{5.1}$$

where n is the number of discharge cycles per year and h is the discharge time, in hours. This approach is adopted in Zakeri and Syri [205]. Another option for calculating the LCOE is offered in Schoenung and Hassenzahl [206], as an alternative in the event that the values of n and/or h_d are not known. Such an alternative is formulated as follows:

$$LCOE = 100\frac{C_{LC}}{h_d \times d}, \tag{5.2}$$

where h_d is the number of operating hours per day and d is the number of operating days per year.

The rest of the chapter introduces a model for the economic assessment of storage technologies, including the metrics presented above. The model will be also tested by discussing a numerical example.

5.2 A Cost Model for Storage Technologies

This section presents a cost model for energy storage technologies. As previously mentioned, cost models usually consider capital, operation and maintenance, and replacement, as well as disposal- and recycling-related costs as the main life-cycle cost components. Such an approach is adopted and discussed in several works (e.g., [93, 205, 206]). The model presented in this section is just the result of a synthetic analysis of the above-mentioned works.

First, tackling the model description, the "annualized life-cycle costs" C_{LC} (in €/kW-yr) are formulated as

$$C_{LC} = C_I + C_{O\&M} + C_R + C_{EoL},\qquad(5.3)$$

where C_I weights the capital costs, which are associated with the acquisition and installation of the system, $C_{O\&M}$ are the operating and maintenance costs during the system's lifetime, the term C_R represents the expected replacement costs due to wear, and C_{EoL} computes the disposal and recycling costs at the end of the lifetime of the system. All these terms are expressed in annualized form, and hence in €/kW-yr. Each of the identified main cost components is described in the following sections.

5.2.1 The Capital Costs

The capital costs considered mainly comprise the cost of the storage container, C_{STO} (e.g., a battery pack), the cost of power conversion systems, C_{PCS} (e.g., electronic power converters), and the cost of the so-called balance of plant components, C_{BoP} (e.g., elements such as power transformers, protection, cooling systems, buildings, and others). Considering all of these components, the estimated capital costs mainly comprise the principal elements that are used to build up the energy storage system (ESS), including not only the storage containers but also all the elements necessary to interface the system with the grid, as well as to ensure its controlled and secure operation. Thus, the total capital costs are computed as follows:

$$C_I^T = (C_{STO} + C_{PCS} + C_{BoP}).\qquad(5.4)$$

Since it represents the initial investment for purchasing and installation, C_I^T is expressed in monetary units. However, such expenditure should be annualized over the time horizon of the project, Y (in yr), and expressed per unit of power rating, P (in kW), for consistency with the rest of the costs contributing to C_{LC} (see Equation (5.3)). The annualization can be performed by affecting total capital costs – per unit of installed power – by the so-called capital recovery factor (CRF) term, which is formulated as follows:

$$CRF = \frac{i(1+i)^Y}{(1+i)^Y - 1},$$ (5.5)

where i is the real discount rate. This term is widely utilized in financial assessments, and concerning costs, can be considered as an annual payment, as a percentage of the amount owed. Expressed in terms of the CRF, the annualized capital costs can be calculated by means of

$$C_I = \frac{C_I^T}{P}CRF.$$ (5.6)

The cost of the storage container is usually proportional to its storage capacity. Thus,

$$C_{STO} = c_e \frac{E}{\mu \times DoD_{max}},$$ (5.7)

where c_e is the specific cost per capacity unit (in €/kWh), E is the storage capacity in kWh, μ is the round-trip energy efficiency – which is given by the quotient between the electricity output at the point of connection of the system and the electricity input – and DoD_{max} is the maximum depth of discharge (DoD) of the technology. The latter is added to consider any possible unusable capacity in storage containers. For instance, such capacity can be related to the remaining energy stored in battery cells at the point at which the minimum operating voltage for the attached power converters is reached, or the remaining capacity in flywheels when the minimum operating rotational speeds for the system are reached.

The cost of the attached power conversion systems is computed as

$$C_{PCS} = c_p P,$$ (5.8)

where the coefficient c_p is the specific cost per power unit (in €/kW), and P is the rated power of the system. The cost C_{PCS} represents, for instance, the cost associated with the electronic power converters managing the battery packs; the cost of the power converters and the electrical machine regulating the energy stored in a rotating disk in flywheel-based systems; and the cost of the turbine and the electrical generator, converting the gravitational potential energy of the water in pumped hydroelectric storage (PHS) reservoirs into electrical energy.

Finally, the balance of plant costs, C_{BoP}, refer to the costs associated with elements such as power transformers, protection, and other equipment for grid connection and integration, as well as ancillary systems such as cooling, monitoring, and environmental isolation. According to Schoenung and Hassenzahl [206], such costs can be expressed with reference to the power rating of the system, and thus in €/kW, and they are highly dependent on the size of the installation. For instance, for bulk storage systems such as PHS, the balance of plant costs are quite important, since dedicated buildings for control and monitoring – as well as electrical equipment for grid connection and integration, such as transformers and even transmission lines – are needed. On the other hand, medium and relatively small technologies such as battery banks or storage for power-quality issues can be included in already existing substations or generating installations, thus minimizing the capital costs in a dedicated balance of plant equipment for the ESS. The total balance of plant costs can be estimated as follows:

$$C_{BoP} = c_{BoP}P, \tag{5.9}$$

5.2.2   Operating and Maintenance Costs

The operating and maintenance costs can be divided into fixed and variable categories. The fixed costs are those that do not depend on the usage of the system throughout its life span. For instance, the fixed costs can be those derived from the contracting of maintenance services for ESSs.

Annualized fixed operating and maintenance costs, $C_{O\&M_F}$, are usually expressed in €/kW-yr. Since they are annualized, they are affected by the annualizing factor L_m, which can be formulated as follows:

$$L_m = CRF \sum_{x=1}^{x=Y} \frac{(1 + d_{cf})^x}{(1 + i)^x}, \tag{5.10}$$

where d_{cf} is the rate of variation of the operating and maintenance fixed costs. Thus, $C_{O\&M_F}$ can be formulated as follows:

$$C_{O\&M_F} = c_f L_m, \tag{5.11}$$

where the term c_f (in €/kW-yr) is the operating and maintenance fixed costs coefficient.

For just short-term ESSs, another important fixed cost is that of purchasing electrical energy to compensate self-discharge losses while providing power quality services. For most of the time, these systems are in the standby state, so they actually exchange electrical energy with the rest of the power system to which they are connected for very short periods of time; for example, in the event of a mains failure while building up "uninterruptible power supply" devices. Thus, continuous provision of energy is necessary to maintain the required SoC. This cost is not considered for mid-term or long-term ESSs, since they are not supposed to be in steady state conditions for the majority of the time, but injecting and absorbing energy to or from the external grid over periods of hours. For such storage installations, the losses due to self-discharge are included in the round-trip efficiency, μ, thus affecting the variable operating and maintenance costs. Ultimately, for short-term ESSs, the annualized fixed operating and maintenance costs can be formulated as follows:

$$C_{O\&M_F} = c_f L_m + \frac{c_{el}}{100} \frac{d}{24} v L_{el}, \tag{5.12}$$

where c_{el} is the electricity cost in c€/kWh, d is the number of operational days per year, and v is the daily self-discharge ratio, expressed per unit of the energy storage capacity. Finally, L_{el} weights the annualizing factor for the cost associated with the electricity purchased, and can be formulated as follows:

$$L_{el} = \mathrm{CRF} \sum_{x=1}^{x=Y} \frac{(1+e)^x}{(1+i)^x}, \tag{5.13}$$

where e is the yearly variation rate for the electricity price.

The variable operating and maintenance costs are those that depend on the usage of the storage system throughout its life span. Apart from the costs of possible eventualities of various kinds, such costs mostly represent the economic expenditure of purchasing electrical energy (and also natural gas for compressed air energy storage, CAES) over the life span of the system. For the sake of clarity, the cost of the electrical energy for compensating parasitic or self-discharge losses is not included here. Instead, within the variable operating and maintenance costs, we consider the energy consumed by ESSs while being actively charged and discharged to provide different services, such as load following or peak shaving.

The variable operating and maintenance costs, $C_{O\&M_V}$, are expressed in €/kW-yr. Just including the cost for purchasing electrical energy, $C_{O\&M_V}$ can be computed as follows:

$$C_{O\&M_V} = \frac{c_{el}}{100\mu} n \times h \times L_{el} = \frac{c_{el}}{100\mu} h_d \times d \times L_{el}, \tag{5.14}$$

where L_{el} is the annualizing factor for the electricity price c_{el}. As can be noted, the terms n and h can be replaced by h_d and d.

CAES systems also consume natural gas apart from electricity, so this should be reflected as a variable operating and maintenance cost. This results in

$$C_{O\&M_v} = \left(\frac{c_{el}}{100\mu} L_{el} + \frac{c_{gas}}{10^6} r_{gas} L_{gas} \right) n \times h = \left(\frac{c_{el}}{100\mu} L_{el} + \frac{c_{gas}}{10^6} r_{gas} L_{gas} \right) h_d \times d, \tag{5.15}$$

where c_{gas} is the cost per gas unit, in €/GJ, r_{gas} is the gas consumption rate per unit of electrical energy provided by the plant, in MJ/kWh, and L_{gas} is the annualizing factor for the gas price, which can be formulated as

$$L_{gas} = \text{CRF} \sum_{x=1}^{x=Y} \frac{(1+b)^x}{(1+i)^x}, \tag{5.16}$$

where b is the inflation rate for the gas price.

5.2.3 Replacement Costs

The components building up an ESS suffer from aging effects to an extent that depends on the required service level and the type of technology. For instance, the expected lifetime of batteries depends on the applied DoD and the discharge rate, the operating temperature, and the charging method, as shown in Figure 5.2.

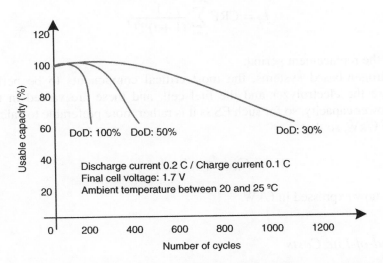

Figure 5.2 The relationship between the lifetime of a lead–acid battery, expressed in terms of cyclability, and the DoD. *Source:* Adapted from Power-Sonic Corporation, 2015 [207].

Since the lifetime of the majority of storage containers (e.g., all battery types) is usually less than the time horizon for the project, replacement costs will be incurred, and these are foreseen by setting up a sinking fund. The total amount of economic resources is usually proportional to the energy storage capacity of the system, and hence expressed in terms of €/kWh. This is because the parts to be replaced during the project time horizon are usually those related to the storage containers (e.g., battery cells). Conversely, the lifetimes of the power conversion systems and the balance of plant equipment are not so critical and could cover most of the time horizon of the project.

Replacement costs could be spread out over each unit of energy output from the ESS. However, such an approach is better adopted while formulating a dynamic model, hence taking into account the energy charged and discharged at each time step over the system's life span. For the purposes of this chapter, though, annualized replacement costs (in €/kW-yr) are estimated by a simple formula, as

$$C_R = \frac{c_r \times h}{\mu} k \times L_r, \tag{5.17}$$

where c_r is the replacement cost coefficient, in €/kWh. This cost is usually quite similar to the specific cost per storage capacity, c_e. The term μ is the round-trip efficiency of the system, h is the number of hours it takes to discharge the storage container in the expected operating conditions, and k is the number of replacements during the project's lifetime. Finally, the annualizing factor L_r is formulated as

$$L_r = \text{CRF} \sum_{k=1}^{x=k} \frac{1}{(1+i)^{r \times \gamma}}, \tag{5.18}$$

where γ is the replacement period.

For hydrogen-based systems, the most critical components to be periodically replaced are the electrolyzer and the fuel cell, and these are valued in terms of installed power capacity, so for such ESSs it is rather more preferable to calculate C_R in terms of €/kW, so that

$$C_R = c_r \times k \times L_r, \tag{5.19}$$

where c_r is now expressed in €/kW.

5.2.4 End-of-Life Costs

To complement the life-cycle cost model, it is important to consider the costs associated with the disposal and recycling of the technology at the end of the system's life span. Energy storage systems use expensive metals, synthetic and organic materials, as well as rare earths, so the environmental impact and economic expenditures associated with their treatment at the end of life of the system should not be neglected.

However, recycling is not a widely implemented practice in the field of storage as yet. With the aim of promoting this practice, in 2006 the European Commission published Directive 2066/66/EU, under which Member States became committed to collecting 25% of the total amount of batteries consumed in each country and in each year by September 2012, and 45% by September 2016. What is important with regard to this directive for storage technology operators is that technology manufacturers are supposed to incur the expenses related to the collection, treatment, and recycling of batteries. Thus, such costs may not be considered for storage operators while evaluating the techno-economic feasibility assessment of the project with regard to installing and operating a storage system. However, they should be considered quantitatively or qualitatively – as for several other aspects, such as social considerations – so as not only to make the decision about technology selection on the basis of economic criteria, but also to adopt a holistic point of view. Apart from the costs associated with the recycling of technology, other unavoidable costs at the end of the project are those related to dismantling the system. Such costs can be especially important in bulk ESSs such as CAES and PHS.

It is not straightforward to determine realistic cost estimates for the aforementioned concepts. In any case, such costs should be integrated into the present cost model in annualized form and in terms of the power capacity of the installation, and hence in €/kW-yr, yielding the following mathematical expression:

$$C_{EoL} = c_{EoL} L_m, \tag{5.20}$$

where c_{EoL} the end-of-life cost coefficient (in €/kW) and CRF is the previously presented annualizing factor.

5.2.5 The Synthesis of a Cost Model

To sum up, the previously described equations for the cost model are succinctly presented in what follows.

Data:

Time horizon of the project	Y	(yr)
Specific cost per storage capacity	c_e	(€/kWh)
Specific cost per power capacity	c_p	(€/kW)
Fixed O&M cost coefficient	c_f	(€/kW)
Replacement cost coefficient	c_r	(€/kWh); (€/kW) for HESS

End-of-life cost coefficient	c_{EoL}	(€/kW)
Cyclability	n_c	(cycles)
Electricity cost	c_{el}	(c€/kWh)
Natural gas cost	c_{gas}	(€/GJ)
Gas consumption rate	r_{gas}	(MJ/kWh)
Round-trip efficiency	μ	(kWh$_{out}$/kWh$_{in}$)
Daily self-discharge ratio	v	(pu)
Number of discharge cycles per year	n	(–)
Discharge time	h	(h)
Operating hours per day	h_d	(h)
Yearly operating days	d	(d)
Number of replacements	k	(–)
Maximum DoD	DoD_{max}	(pu)
Nominal discount rate	i	(pu)
Variation rate for O&M fixed costs	d_{cf}	(pu)
Variation rate for electricity price	e	(pu)
Variation rate for gas price	b	(pu)

Equations:

Annualized life-cycle costs	$C_{LC} = C_I + C_{O\&M} + C_R + C_{EoL}$	(€/kW-yr)
Total capital costs	$C_I^T = C_{STO} + C_{PCS} + C_{BoP}$	(€)
Annualized capital costs	$C_I = \frac{C_I^T}{YP}$	(€/kW-yr)
Total cost of storage container	$C_{STO} = c_e \frac{E}{\mu DoD_{max}}$	(€)
Total cost of PCS	$C_{PCS} = c_p P$	(€)
Total cost of BoP	$C_{BoP} = c_{BoP} P$	(€)
Annualized fixed O&M costs	$C_{O\&M_F} = c_f L_m$	(€/kW-yr)
Annualized fixed O&M (short-term)	$C_{O\&M_F} = c_f L_m + \frac{c_{el}}{100} \frac{d}{24} v L_e$	(€/kW-yr)
Annualized variable O&M costs (I)	$C_{O\&M_V} = \frac{c_{el}}{100\mu} n \times h \times L_{el}$	(€/kW-yr)
Annualized variable O&M costs (II)	$C_{O\&M_V} = \frac{c_{el}}{100\mu} h_d \times d \times L_{el}$	(€/kW-yr)
Annualized variable O&M costs CAES (I)	$C_{O\&M_V} = \left(\frac{c_{el}}{100\mu} L_{el} + \frac{c_{gas}}{10^3} r_{gas} L_{gas} \right) n \times h$	(€/kW-yr)
Annualized variable O&M costs CAES (II)	$C_{O\&M_V} = \left(\frac{c_{el}}{100\mu} L_{el} + \frac{c_{gas}}{10^3} r_{gas} L_{gas} \right) h_d \times d$	(€/kW-yr)
Annualized replacement costs	$C_R = \frac{c_r \times h}{\mu} k \times L_r$	(€/kW-yr)
Annualized replacement costs (HESS)	$C_R = c_r \times k \times L_r$	(€/kW-yr)

Annualized end-of-life costs	$C_{EoL} = CRF \times c_{EoL}$	(€/kW-yr)
Annualizing factor for O&M fixed costs	$L_m = CRF \sum\limits_{x=1}^{x=Y} \frac{(1+d_{cf})^x}{(1+i)^x}$	(pu)
Annualizing factor for electricity	$L_{el} = CRF \sum\limits_{x=1}^{x=Y} \frac{(1+e)^x}{(1+i)^x}$	(pu)
Annualizing factor for natural gas	$L_{gas} = CRF \sum\limits_{x=1}^{x=Y} \frac{(1+b)^x}{(1+i)^x}$	(pu)
Levelized cost of energy (I)	$LCOE = 100\frac{C_{LC}}{n \times h}$	(c€/kWh)
Levelized cost of energy (II)	$LCOE = 100\frac{C_{LC}}{h_d \times d}$	(c€/kWh)

5.3 An Example of an Application

This section presents an example of an application of the cost model introduced in this chapter.

Considering the principal characteristics of each of the storage technologies presented in Chapter 4 (e.g., batteries, flow batteries, PHS, CAES, flywheels, and so on), there is no doubt that each one is best suited for a different purpose. For instance, the fast response and high cyclability of supercapacitors are quite pertinent for providing power quality services, such as flicker filtering and smoothing of the output of renewables. However, their limited energy capacity makes them useless for peak shaving purposes; conversely, PHS installations and CAES are well suited for that service. Accordingly, the ESSs presented in Chapter 4 can be classified into three main categories, addressing their suitability for providing one type of service or another. This facilitates the evaluation of results and supports their credibility. In particular, the three categories are defined below in terms of the required energy and power capacity for ESSs while providing particular services:

- **Long-term, high-power storage**. In this category, ESSs are rated at 100 MW/ 600 MWh, so they are able to provide huge amounts of power continuously, for up to 6 h. Such a high capacity defines ESSs as suitable, for instance, to time-shift the output of renewables when convenient, according to market and/or technical power system constraints, and even to store energy seasonally. For a description of the applications that ESSs can provide in the electric power system, see Chapter 8. The storage technologies included in this category are PHS and CAES.
- **Mid-term, mid-power storage**. In this category, ESSs are rated at 10 MW/10 MWh. The capability to inject or absorb multi-megawatt power for up to 1 h defines the ESSs included in this category as suitable for helping renewables to meet their output targets, and to help the network operator to continuously ensure the required

balance between generation and demand, among other applications. The storage technologies included in this second category are batteries, flow batteries, and hydrogen-based technologies. Although PHS and CAES are also well suited for providing the above-mentioned services, they are not included in this category because their ratings normally exceed those prescribed here. Similarly, flywheel-based plants, although well suited for providing a multi-megawatt output for a relatively long period of time (for up to 30 minutes [169]), are not included in this category, because they normally are best suited to providing services related to power quality.

- **Short-term, low-power storage**. To complete the classification, this third category concerns those ESSs rated around 1 MW/3 kWh. The key characteristics of the ESSs included here are their high cyclability, their short time response, and their limited energy capacity. In fact, the capability to regulate power continuously for less than a minute limits their applicability to services related to power quality (e.g., power smoothing of the output of renewables, flicker filtering, support under grid faults, and so on). Accordingly, the technologies included here are principally flywheels, supercapacitors, and superconducting magnetic energy storage (SMES). Despite the fact that advanced devices such as lithium-based batteries and flow batteries could provide the services considered in this category, they are left out because their energy storage ratings usually exceed those prescribed here.

This section presents the results obtained from the application of the cost model concerning the ratings for ESSs as defined in the above classification. First, though, and for evaluation of the cost model, the required technical data are presented and discussed in the following subsection.

5.3.1   The Collection of Data for Evaluation of the Cost Model

This section offers a set of cost data for different types of storage technologies. The cost data are utilized to perform a rough economic analysis of various ESSs while providing particular services for the electric power system, and hence building up stationary storage facilities. The data used are summarized in Tables 5.1, 5.2, 5.3, and 5.4, which have been deduced from the tables summarizing storage technology characteristics in Chapter 4.

In addition, Table 5.1 lists the adopted values for the project time horizon, the electricity price, the real discount rate, the electricity cost variation rate, and O&M fixed costs variation rate. These cost data are common to all ESSs.

The parameters for evaluation of the cost model with regard to long-term ESSs are summarized in Table 5.2. As can be noted, for such installations, we consider the application of peak shaving (for more information, see Chapter 8). Briefly, this means that ESSs will be discharged once per day during peak hours, so as to avoid

Table 5.1 Cost data common to all ESSs.

Parameter	Value
Y (yr)	20
c_{el} (c€/kWh)	5
i (pu)	0.085
e (pu)	0.010
d_{cf} (pu)	0.0

the activation of conventional peak power plants based on fossil fuels or gas. In doing that, the ESSs will be discharged $n = 365$ times per year. Ultimately, the total number of discharge cycles during the $Y = 20$ yr project lifetime will not surpass the cyclability of the storage plants, so the replacement costs will be minimized. Analyzing the parameters in Table 5.2, it is important to note that the specific cost per energy unit is relatively inexpensive. On the other hand, the specific cost per power unit is much higher, reflecting the high capital costs of electrical equipment, turbines, heat exchangers (for CAES systems), and so on.

Table 5.2 Data for long-term ESSs.

	PHS	CAES
c_e (€/kWh)	20	5
c_p (€/kW)	600	400
c_{BoP} (€/kW)	–	–
c_f (€/kW)	8.5	13
c_r (€/kWh)	–	–
k (–)	0	0
n_c (–)	5×10^4	3×10^4
γ	50	40
μ (pu)	0.75	0.71
v (pu)	1×10^{-5}	1×10^{-5}
n (–)	365	365
h (h)	6	6
h_d (h)	6	6
d (d)	365	365
DoD_{max} (pu)	0.9	0.6
c_{gas} (€/GJ)	Not applicable	9.97
r_{gas} (MJ/kWh)	Not applicable	4.43
b (pu)	Not applicable	0.03

Table 5.3 Data for mid-term ESSs.

	HESS	VRB	ZBB	NaS	Lead–acid	Ni–Cd	Li-ion
c_e (€/kWh)	10	600	500	350	150	780	750
c_p (€/kW)	1180	150	150	150	150	150	150
c_{BoP} (€/kW)	0	50	50	50	50	50	50
c_f (€/kW)	23.6	5	5	5	5	5	5
c_r (€/kWh)	413 €/kW	300	250	350	150	780	750
k (–)	1	1	2	2	3	2	2
n_c (–)	1×10^4	1×10^4	2.5×10^3	2.5×10^3	2×10^3	3×10^3	3×10^3
γ	10	10	7	7	5	7	7
μ (pu)	0.42	0.85	0.74	0.82	0.75	0.72	0.90
v (pu)	1×10^{-5}	1×10^{-5}	1×10^{-5}	1×10^{-5}	2×10^{-3}	3×10^{-3}	3×10^{-2}
n (–)	365	365	365	365	365	365	365
h (h)	1	1	1	1	1	1	1
h_d (h)	1	1	1	1	1	1	1
d (d)	365	365	365	365	365	365	365
DoD_{max}	1	1	0.8	0.8	0.8	0.8	0.8

Table 5.4 Data for short-term ESSs.

	SMES	FESS	SCESS
c_e (€/kWh)	72×10^3	90×10^3	57×10^3
c_p (€/kW)	300	150	150
c_{BoP} (€/kW)	150	125	25
c_f (€/kW)	5	5	5
c_r (€/kWh)	72×10^3	90×10^3	57×10^3
k (–)	1	0	1
n_c (–)	5×10^5	1×10^6	5×10^5
γ	10	20	10
μ (pu)	0.90	0.90	0.85
v (pu)	0.1	1	0.2
n (–)	5×10^4	5×10^4	5×10^4
h (h)	0.003	0.003	0.003
h_d (h)	0.38	0.38	0.38
d (d)	365	365	365
DoD_{max} (pu)	0.8	0.8	0.75

Table 5.3 summarizes the parameters for mid-term ESSs. As for long-term storage, we consider the application of peak shaving, so that ESSs are fully discharged once per day until their cyclability is exhausted. Then, the storage containers (i.e., the battery cell stacks) should be replaced, incurring important replacement costs. The lifetimes for power conversion systems are usually much longer than for storage containers.

Further analyzing the data in Table 5.3, it is worth noting that all battery types feature the same specific cost per power unit, of $c_p = 150$ €/kW. This is because similar technology for the power conversion systems – that is, the electronic power converter interfacing the main grid with the battery pack – is intended. This specific cost is obtained from the supplier's quotations.

For hydrogen-based systems, the term c_p is much higher. This is representative of the cost associated with the electrolyzer and fuel cell of the system (for more information about the system topology, see Chapter 4). The value presented here was derived from the National Renewable Energy Laboratory (NREL) [208, 209], which reports data on capital costs for a PEMFC electrolyzer rated at 2.33 MW, also including the associated power electronics. While c_p makes a major contribution to the capital costs of the system, the cost per energy storage capacity, c_e, is quite inexpensive, mostly representing the cost of metallic hydrogen gas containers.

Looking now at the balance of plant costs, c_{BoP}, these are assumed to be approximately around one third of the cost associated with the power conversion components for all systems. Moreover, low operating and maintenance fixed costs are assumed. The value indicated in the table for secondary and flow batteries is obtained by referring the salary of a technician (around 50 k€/yr) to the installed capacity (i.e., 10 MW).

Replacement costs are directly related to the cost of storage containers for flow and secondary batteries, but not for hydrogen-based systems. For secondary batteries, the replacement costs equal the specific cost per energy capacity unit, c_e. For flow batteries, though, the replacement costs are just around 50% of c_e, as indicated by the ZBB manufacturer Redflow [113]. This difference between c_e and c_r represents the fact that c_e includes the cost of both cell stacks and electrolyte containers, but it is just the cell stacks that have to be periodically replaced. As noted in Section 5.2.3, the replacement costs for hydrogen-based systems are expressed in €/kW. To summarize, note that the number of replacements k for all technologies is computed by comparing the cyclability of cell stacks with the yearly discharge cycles.

Finally, it is important to note that the indicated DoD is in accordance with the assumed cyclability. For instance, and according to the manufacturer's information, lithium-ion (Li-ion) batteries can be discharged just a few thousand times, down to 20% of their storage capacity, but the technology allows up to 400 000 cycles to be performed at 2.2% DoD. Moreover, it is important to note that although some technologies can be fully discharged without incurring cell damage (as for flow batteries, and Li-ion and Ni–Cd batteries), the minimum state of charge for the operability of the power converter has been limited.

The parameters for short-term storage are presented in Table 5.4. Here, the ESSs are configuring uninterruptible power supply systems. Thus, they are not frequently fully discharged, but only in the event of mains failures. In normal operating conditions, the ESSs are continuously charged and discharged over a short period of time to filter out harmonics, voltage flicker, and other issues related to power quality, so as to protect the equipment of sensible consumers. The same operating principle of active filtering can be applied when included in a renewable-based generating plant, looking at its grid connection.

Briefly, it is important to note that the balance of plant costs are much higher for SMES and FESS than for secondary batteries because of the need to include vacuum and cryogenic systems. On the other hand, the balance of plant costs for ultracapacitors are considered lower than for secondary batteries (around 50 €/kW; see Table 5.3) because of the simplicity of the state estimators [210] (the SoC is directly proportional to the voltage in the supercapacitor cell). Energy Management Systems (EMSs) for batteries are much more complicated and expensive.

Moreover, it is convenient to note that the calculation of the number of equivalent yearly full discharge cycles, $n = 50\,000$, was performed by dividing the cyclability of flywheels (1 million cycles) by its lifetime (20 yr). Considering the resultant value, as well as addressing the life span of SMES and supercapacitors, it has been concluded that these storage containers are to be replaced once during the project time horizon.

5.3.2   Analysis of the Results

This section presents the results obtained from the evaluation of the cost model while applying the parameters introduced previously. The comparison of technologies is in accordance with the proposed classification shown above; that is, long-, mid-, and short-term ESSs.

5.3.2.1   Results for Long-Term Storage Installations

Here, the evaluation of the results is concentrated on PHS and CAES systems. Figure 5.3 compares the life-cycle cost breakdowns for both technologies.

As can be noted, the annualized cost per installed power capacity is presumably lower for PHS than for CAES systems. Despite the fact that the capital costs are higher for PHS, the electricity and fuel costs incurred for CAES systems boost the annualized costs for this technology considerably.

Translating these results into an equivalent LCOE (see Table 5.5), it can be derived that the cost of both technologies, under the assumptions of the present study and while performing the previously indicated service, turn out to be quite similar to the cost of conventional fossil-fuel power plants (around 12 c€/kWh) [94]). The low LCOE of PHS explains its extensive deployment in the power system.

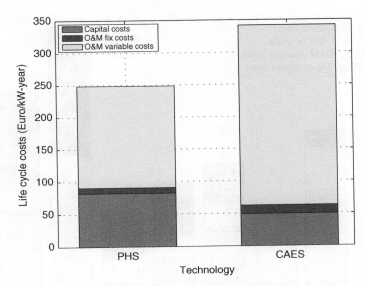

Figure 5.3 The annualized life-cycle costs for long-term ESSs. The systems are rated at 100 MW/600 MWh, and are deeply discharged once per day throughout the project horizon.

5.3.2.2 Results for Mid-Term Storage Systems

Here, the evaluation of the results in concentrated on conventional batteries, flow batteries, and hydrogen-based systems. Figure 5.4 compares the life-cycle cost breakdowns for these technologies.

As can be observed, the capital costs of secondary and flow batteries greatly impact the system's life-cycle costs. Furthermore, and because of the limited cyclability of storage containers, they are to be periodically replaced, thus adding important expenditure.

The presented costs breakdown also indicates that economic resources for purchasing electricity (affecting the O&M variable costs) are less important than for the case of bulk storage systems, as presented in the previous section. This is because of the specificities of the final application that the storage systems are providing here (they become deeply discharged once per day, and the discharge time is just 1 h). However, it is worth noting that the expenditure for purchasing electricity for hydrogen-based

Table 5.5 Data for long-term ESSs.

	PHS	CAES
LCOE (c€/kWh)	11.4	15.6

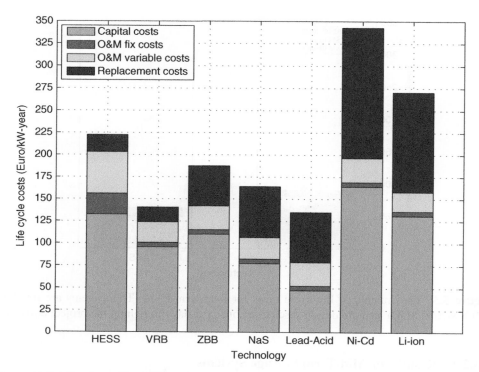

Figure 5.4　The annualized life-cycle costs for long-term storage systems. The systems are rated at 10 MW/10 MWh, and are deeply discharged once per day throughout the project horizon.

systems is comparatively higher than for batteries and flow batteries. This is because of the relatively low round-trip efficiency (around 42%) of regenerative fuel cells, in comparison with the other technologies.

Evaluating the global picture, we can conclude that the low capital costs of lead–acid batteries define this technology as the one with the lowest life-cycle costs. However, the high cyclability, life span, and low replacement costs of flow batteries (especially for VRBs), configure them as promising alternatives for mid-term, multi-megawatt systems. However, it is important to note that the low specific energy of flow batteries mostly restricts their usage to stationary applications (so they are not quite suitable for application in the field of electromobility, for instance).

Finally, we just remark that the still high capital costs of Li-ion batteries greatly constrains their applicability, especially when configuring multi-megawatt systems. Currently, there are other options that seem economically preferable. However, with regard to the high performance and intensive research and development activities for lithium batteries, the costs indicated here will presumably begin to diminish in the near future.

Table 5.6 Data for mid-term ESSs.

	HESS	VRB	ZBB	NaS	Lead–acid	Ni–Cd	Li-ion
LCOE (c€/kWh)	60.9	38.5	51.4	44.9	**36.9**	**93.9**	74.1

If we now evaluate the LCOE indices (see Table 5.6), it is clear that lead–acid batteries are viewed as the most economical alternative. Again, the results highlight the necessity of reducing the costs for Li-ion (and also Ni–Cd) batteries.

5.3.2.3 Results for Short-Term Storage Systems

Here, the evaluation of the results is concentrated on SMES, and on flywheel-based and supercapacitor-based storage systems. Figure 5.5 compares the life-cycle cost breakdowns for these technologies.

As can be noted, the predominant cost concept contribution for all three technologies is the capital cost. Furthermore, it can be derived that SMES life-cycle costs are

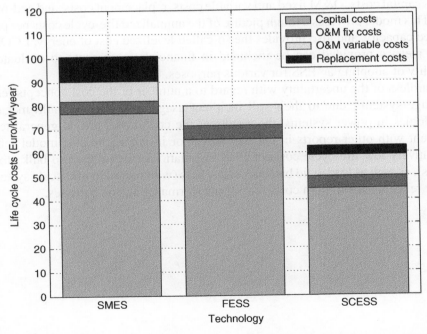

Figure 5.5 The annualized life-cycle costs for short-term storage systems. The systems are rated at 1 MW/0.003 MWh, taking 50 000 equivalent discharge cycles per year into consideration.

Table 5.7   Data for short-term ESSs.

	SMES	FESS	SCESS
LCOE (c€/kWh)	**72.5**	57.5	45.1

presumably the highest amongst the presented alternatives. However, since SMES is a young technology and there is a lack of literature reporting real system performance, one should consider the high degree of uncertainty in the cost of SMES.

Moreover, it is important to note that the life-cycle costs for flywheel-based systems are reduced by the fact that no replacement costs are incurred during the lifetime of the project. If they were, the costs would be increased slightly.

Finally, LCOE values for all three technologies are presented in Table 5.7, which also depicting similar costs for flywheel- and supercapacitor-based systems.

5.4   Conclusions

This chapter has presented a cost model for ESSs, which includes different cost terms such as capital costs, O&M fixed and variable costs, replacement costs, and end-of-life costs. This model provides a rough picture of the annualized life-cycle costs per power installed capacity, and also provides the so-called levelized cost of energy, LCOE.

The metrics provided can be included in financial assessments to evaluate the suitability of adopting an ESS for various purposes.

Regardless of the uncertainty with regard to a number of the cost concepts related to storage systems, due to their still incipient usage as stationary multi-megawatt technologies in power systems, the results of the cost model are, in general terms, consistent with other reports in the literature. For instance, the cost model reflects the relatively low life-cycle costs of PHS installations (with regard to bulk storage systems), as well as lead–acid batteries and VRBs (with regard to mid-scale systems). Also, it reflects the still high costs for high-performance Li-ion batteries.

6

Modeling, Control, and Simulation

6.1 Introduction

The study of the electric power system, together with the field of renewables and also energy storage technologies, is a multidisciplinary field of knowledge, involving electrical, mechanical, and chemical engineering, among others. Hence research in this field requires a broad and deep grasp of several aspects of knowledge, which can be difficult and can compromise success and advancement. Additionally, research in this field is further affected by the fact that the subject matter is hardly reproducible, for a number of technical and economic reasons. For instance, a high-voltage power system is, for the majority of researchers, impossible to reproduce, even at laboratory scale, due to the high cost of the technology. Few research institutions could develop a test site including various kinds of real generators (e.g., fuel-based, photovoltaic, wind turbines, etc.), differing types of load (e.g., passive and active), various types of storage technology, and all the electrical equipment needed to build up a power distribution layout – for example, power lines, transformers, and protection systems, among others. Furthermore, a generator or a power line, once installed in the power system, can hardly be tested, since it is actually supplying power to consumers, who would not be willing to suffer the inconvenience of eventual interruptions to their power supply.

The above-mentioned issues are some of the reasons that motivate the development of simulation platforms. These allow us to represent, on the basis of mathematics, real systems such as actual megawatt generators, high-voltage power lines, and storage technologies, among others. Since they are based on mathematics, simulation platforms allow us to test the behavior of the above-mentioned systems in different environments and operating conditions, as if they were actually installed and running, but at practically no cost (no real equipment has to be purchased) and avoiding any inconvenience to final users – for example, the security of supply to final users is not affected by the fact that a new power line has to be tested under short-circuit conditions – among several other reasons.

Energy Storage in Power Systems, First Edition. Francisco Díaz-González, Andreas Sumper and Oriol Gomis-Bellmunt.
© 2016 John Wiley & Sons, Ltd. Published 2016 by John Wiley & Sons, Ltd.

In this way, modeling and simulation are key factors for studies related to power systems and storage technologies, and as such, they are the main subjects of this chapter. In particular, the objectives of the chapter are twofold: (i) to provide an initial idea about how to model energy storage systems (ESSs), depending on the objectives of the simulation; and (ii) to present dynamic models, based on electrical equations, for different storage technologies such as batteries, supercapacitors, and flywheels, as well as for the attached power conversion systems. In addition, to effectively show charging and discharging profiles, a broad range of the control schemes utilized for power conversion systems are also presented and tuned.

6.2 Modeling of Storage Technologies: A General Approach Orientated to Simulation Objectives

There are various ways to model energy storage technologies. For instance, one can consider the storage system as a simple black box and apply simple mathematics obtain a few steady-state magnitudes from the inputs or setpoints indicated. But one can also model each of the components of a storage system in order to be able to evaluate internal energy and mass fluxes that determine transient and steady-state outputs of differing natures. Therefore, ultimately, models must be adapted to the objectives of the particular study for which they are developed. The selection of the type of model that should be applied in each case must take several aspects into account, such as:

- **The required input data and the output signals of the model**. These determine, for instance, the connectivity of the model with other systems.
- **The required simulation time steps**. These determine and constrain the events to be represented by the simulation. For instance, small time steps (in the order of microseconds) are usually needed to represent fast electrical transients in voltage and current waveforms due to short circuits in a power line.
- **The required simulation time horizons**. For instance, long time horizons are needed to evaluate the behaviour of an ESS on a daily basis, but they are not required to evaluate fast electromagnetic transients.
- **Other computational and functional constraints**. The complexity of the models, with regard to the calculus to be carried out in each simulation time step, may be restricted to computational and functional aspects. For instance, sometimes models should be executed on a real-time basis (e.g., in the case of hardware-in-the-loop simulators [211]) and this means that the outputs of the simulation are continuously being used by other computational environments and for other purposes. In such cases, it is important for a simulation solver to converge to a solution in a very restrictive and usually narrow time interval, and so the complexity of the model should be adapted accordingly.

With regard to studies related to power systems (also including here the field of renewable generation and ESSs), one can roughly establish a relationship between the characteristics of the models to be used and the objectives of the study. In particular, three types of model are defined here: *simplified*, *averaged*, and *detailed* models.

Simplified models are characterized by the fact that they are based on simple mathematics, allowing the computation of mainly steady or averaged energy and mass flows, given some operational conditions. In this manner, they provide a snapshot of the state of the system, allowing the evaluation of averaged figures from a systems perspective. These models are very useful in the sense that they can even be easily implemented using spreadsheets, thus enabling simplified techno-economic parametric studies, among others.

Averaged models are intended here as those characterized by the fact that they allow the development – with a good degree of complexity and fine detail – of both transient and steady state calculations. With regard to those utilized in studies related to power systems, such models are made up of differential equations, enabling the representation of electrical and mechanical transients. For instance, an averaged model for a wind turbine could be one that represents the electrical transients by means of a fifth-order model of the electrical generator to which the wind turbine rotor is attached [63].

Finally, *detailed models* are intended here as those able to not only reproduce the behaviour of the system from a systems perspective, but also to represent the behavior of its components at various levels (e.g., including mechanical, electrical, and electromagnetic transients) in detail. The capacity to reliably reproduce the electromagnetics of the system is, indeed, one of the main differences between averaged and detailed models, as far as studies relating to power systems are concerned. Such models can be used to test the fast electrical and magnetic dynamics of various technologies under disturbances such as short circuits. They can also be used to represent power quality issues, such as voltage and current harmonic emission, and electromagnetic compatibility. The simulation of detailed models requires a high level of computational effort because of the complexity of the mathematics involved and the very restrictive time steps. For instance, one example of a detailed model is that able to represent the switching of the transistors used to build up a electronic power converter.

The rest of the contents of this chapter discusses averaged models for batteries, supercapacitors, and flywheels, also including their corresponding power conversion systems for control and grid connection. For modeling purposes, all storage systems are intended here to be connected to an AC external grid, which can be of either a medium- or low-voltage type. With regard to medium-voltage networks, the storage installations are connected through a step-down low-frequency transformer, and a set of two power electronic converters, hence building up the power conversion system. The first of these two converters is called the "grid-side converter" (or "GSC") hereinafter and is considered to always be a three-phase two-level inverter, which interfaces the three-phase low-voltage AC external system with the DC voltage at the terminals of an inner power converter. This second power converter, called the "storage-side converter" (or "SSC") hereinafter, interfaces the AC or DC terminals of the storage containers, depending on their type, with the DC side of the GSC.

Moreover, the duty of the GSC is to always keep the voltage on its DC side stable and constant. It is also in charge of regulating the reactive currents exchanged with the network. Conversely, the duty of the SSC is to always actively manage the state of charge of the system.

Thus, to address the common features for each of the power converters of the power conversion system for all storage installations, and to avoid any redundancy in explanation, the contents are organized as follows. First, the modeling and control of the bidirectional three-phase two-level power inverter used to build up the GSC for all kinds of storage systems is explained in Section 6.3. Second, the modeling and control of the storage containers, while attached to their corresponding SSCs, is depicted in Section 6.4. The latter includes dedicated contents for the description of the DC–AC rectifier – flywheel coupling and for the DC–DC converter – battery coupling, as well as for the DC–DC converter – supercapacitor coupling. Finally, Section 6.5 presents some simulation results depicting the performance of the previously introduced models and controllers.

The presented approach is graphically depicted in Figure 6.1. Note that both power converters are interfaced by a DC link, and that the GSC interfaces with the external grid through an inductive filter. These components or parts are also included in the model.

6.3   The Modeling and Control of the Grid-Side Converter

6.3.1   Modeling

As previously noted, the considered topology for the GSC is the three-phase two-level inverter. The utilized power switches are IGBTs. As noted in Figure 6.1, the converter is in charge of ensuring a stable and constant DC voltage, for the proper operation of the SSC. Also, the GSC regulates the reactive currents exchanged with the main AC grid.

The inverter's DC bus is charged and discharged through an inductive filter on the AC side. Considering both the inverter and the external grid as ideal AC voltage sources, the external grid – inductive filter – inverter system resembles the equivalent circuit plotted in Figure 6.2. Note that by adopting this approach, possible high-frequency effects such as harmonics, and diverse phenomena induced by the switching of the IGBTs used to build up the power inverter, are not considered.

For modeling and control design purposes, it is usually preferable not to consider time-varying magnitudes, such as sinusoidal voltages and currents, but steady ones. This greatly simplifies the related mathematics and facilitates evaluation of the results. Thus, in this regard voltages and currents are not treated as seen in a stationary frame of reference, in which they appear as sinusoidal magnitudes that vary with the grid frequency. Instead, they are weighted as seen in a rotating frame of reference with,

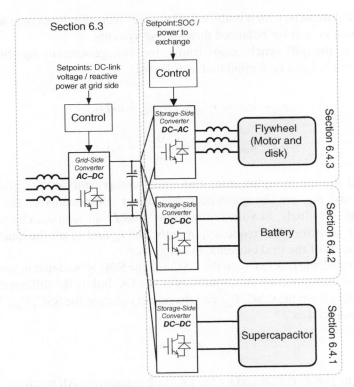

Figure 6.1 The organization of the contents for describing the modeling and control of various kinds of storage system attached to their corresponding power conversion systems.

usually, the grid frequency, such that time-dependent voltages and currents appear to be constant in time – provided that their frequency matches that of the frame of reference. The mathematical transformation that expresses time-varying magnitudes seen in a stationary frame of reference in rotating form, synchronous with the grid frequency, is known as the $qd0$, or Park's transformation [213]. In this frame of reference, the three-phase ABC magnitudes are expressed using three axes: the direct

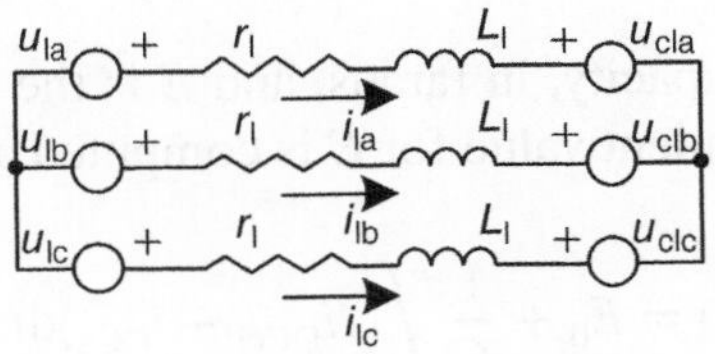

Figure 6.2 The modeling of the GSC circuit. The voltages u_{labc} correspond to the grid-side terminals, while the voltages u_{clabc} are at the converter terminals. *Source:* Díaz-González *et al.*, 2013. Reproduced with permission of Elsevier.

axis; the quadrature axis, which is shifted 90 degrees from the direct one; and the zero sequence, which is zero for balanced three-phase systems.

Expressed in the $qd0$ synchronous frame, the voltage–current equations for the system in Figure 6.2 can be formulated as follows:

$$u_{1q} - u_{clq} = r_1 i_{1q} + L_1 \frac{d}{dt} i_{1q} + \omega_1 L_1 i_{1d}, \tag{6.1}$$

$$u_{1d} - u_{cld} = r_1 i_{1d} + L_1 \frac{d}{dt} i_{1d} - \omega_1 L_1 i_{1q}, \tag{6.2}$$

where u_{1q} and u_{1d} are the quadrature and direct components of the grid voltages, in volts; u_{clq} and u_{cld} are the quadrature and direct components of the voltage at the inverter AC terminals, in volts; r_1, in ohms, and L_1, in henries, characterize the inductive filter; ω_1 is the grid frequency, in rad/s; and i_{1q} and i_{1d} are the quadrature and direct components of the grid currents, in amperes.

Moving forward, the link between the GSC and the SSC is modeled in what follows. First, the power consumed by the capacitors of the DC link is the difference between the power coming from the GSC, P_{cl}, and the power entering the SSC, P_{cs}. This power balance is formulated as

$$P_{cl} - P_{cs} = P_{DC}, \tag{6.3}$$

where P_{DC} is the power consumed by the DC-link capacitors. All terms are expressed in watts. The output, P_{cl}, neglecting internal power losses into the converter, can resemble its input on the AC side, yielding the following expression in terms of $qd0$ voltages and currents:

$$P_{cl} = \left(\frac{3}{2}\right)(u_{cld} i_{cld} + u_{clq} i_{clq}). \tag{6.4}$$

In turn, the power consumed by the capacitors in the DC link is expressed by

$$P_{DC} = \frac{1}{2} C \frac{d}{dt} E^2, \tag{6.5}$$

where C is the equivalent capacity, in farads, and E is the DC-link voltage, in volts. Furthermore, the time-dependent value for E is computed as follows:

$$E(t) = E_0 + \frac{1}{C} \int_0^t (i_{DCcl} - i_{DCs})dt, \tag{6.6}$$

where E_0 is the DC bus voltage at time $t = 0$; i_{DCcl} is the DC current from the GSC; and i_{DCs} is the DC injected current to the SSC circuit.

The set of equations presented so far builds up the model for the GSC, and also includes the electrical circuits interfacing it with the external grid on the AC side, and with the SSC on the DC side. Using these equations, what follows describes the related control system for the converter.

6.3.2   Control

As previously noted, the GSC controller is responsible for keeping the DC-link voltage stable and constant, according to a reference value. Also, it is in charge of regulating the reactive currents exchanged with the external grid at the connection point. An instantaneous power theory-based algorithm [63, 214–216] is adopted to control these magnitudes, yielding the control scheme shown in Figure 6.3.

As shown, the input measurements are the alternating currents flowing from the converter, i_{labc}, the sinusoidal voltages at points of common coupling, u_{labc}, and the DC-link voltage, E. The setpoints are the desired DC-link voltage, E^*, and the direct-axis current component flowing from the converter, i_{ld}^*, which is directly associated with the reactive current, as discussed below. From these measurements and control setpoints, the algorithm provides the setpoints for the sinusoidal voltage desired at the converter terminals, expressed in a stationary frame $u_{cl\alpha\beta}^*$. These voltage setpoints are the inputs to a space vector pulse width modulation (SVPWM) control scheme, which governs the switching of the IGBTs. Since the converter is simply modeled as three-phase voltage sources, according to the scope for averaged simulation models, such a SVPWM control scheme is not implemented.

As presented in Figure 6.3, the control system is mainly composed of two cascading control loops. The inner control loop ensures that the qd currents entering the converter follow the indicated setpoints outputted by the outer control loop, which governs the DC-link voltage dynamics. Both control loops are designed in what follows.

First, however, it is important to clarify that for control design, the $qd0$ magnitudes are to be orientated, so that with regard to the d-axis voltage at the point of common coupling of the system, u_{ld} is zero. If one does this, the formulation results are simplified, as is the tuning of the controllers. The proper angle for expressing the abc alternating magnitudes in a $qd0$ synchronous reference with the network frequency, and orientated in a manner such that $u_{ld} = 0$ by applying Park's transformation, is provided by the phase-locked loop (PLL) module shown in Figure 6.3.

6.3.2.1   The Current Control Loop

This section presents the tuning of the PI controllers for i_{qd} currents. According to Equations (6.1) and (6.2), the q component of the GSC voltage u_{clq} depends on the d-axis component of the current flowing into the grid, i_{ld}. Similarly, the d-axis voltage

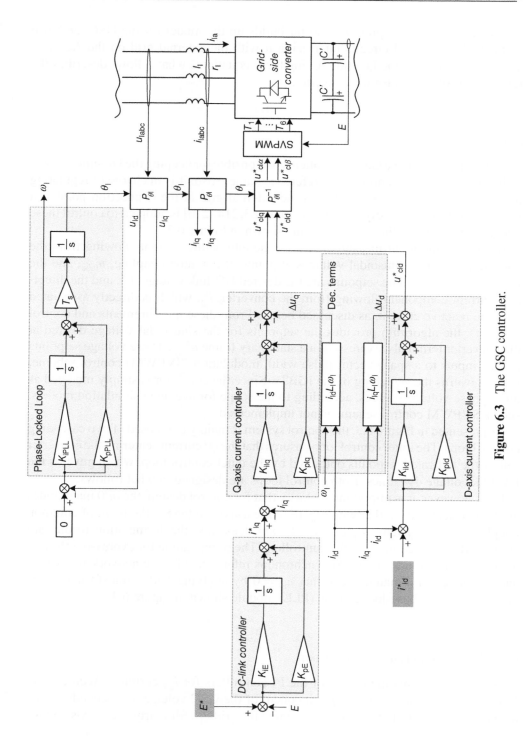

Figure 6.3 The GSC controller.

component depends on the q-axis current component. Accordingly, Equations (6.1) and (6.2) can be rearranged as follows:

$$u_{clq} = -\hat{u}_{clq} + \underbrace{u_{lq} - \omega_1 L_1, i_{ld}}_{\Delta u_q} \tag{6.7}$$

and

$$u_{cld} = -\hat{u}_{cld} + \underbrace{\omega_1 L_1 i_{lq}}_{\Delta u_d}, \tag{6.8}$$

where $\hat{u}_{clq}$ and $\hat{u}_{cld}$ form a decoupled first-order type system $G(s)$. Expressed in the Laplace domain, this takes the following form:

$$i_{lqd} = \underbrace{\begin{pmatrix} \frac{1}{r_1+L_1 s} & 0 \\ 0 & \frac{1}{r_1+L_1 s} \end{pmatrix}}_{G(s)} \hat{u}_{clqd}. \tag{6.9}$$

A linear controller $C(s)$ can now be proposed to regulate the qd currents in terms of the simplified plant or the first-order decoupled transfer function $G(s)$, as indicated in Figure 6.4.

The prescription for controllers tuning into the direct synthesis design method [217] indicates that the controller should be selected so that the resulting closed-loop transfer function $M(s)$ is as predefined by the designer. In this case, $M(s)$ is needed in the form of a first-order type transfer function, dependent on a parameter or time constant λ_1:

$$M(s) = \frac{1}{1 + \lambda_1 s}. \tag{6.10}$$

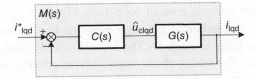

Figure 6.4 The block diagram for design of the current control loop.

Thus, first addressing the q-axis controller, the transfer function for $C(s)$ can be derived by computing

$$C(s) = \frac{M(s)}{G_{11}(s)(1 - M(s))} = \frac{L_1}{\lambda_1} + \frac{r_1}{\lambda_1 s} = K_{plq} + \frac{K_{ilq}}{\lambda_1 s}, \qquad (6.11)$$

where the one–one component of the matrix $G(s)$ is given in Equation (6.9). The transfer function for $C(s)$ corresponds to that for a PI controller, with parameters K_{plq} and K_{ilq}, which are formulated as follows:

$$K_{plq} = \frac{L_1}{\lambda_1} \qquad (6.12)$$

and

$$K_{ilq} = \frac{r_1}{\lambda_1}. \qquad (6.13)$$

Analogously, the parameters for the d-axis current controller can be obtained as follows:

$$K_{pld} = \frac{L_1}{\lambda_1} \qquad (6.14)$$

and

$$K_{ild} = \frac{r_1}{\lambda_1}. \qquad (6.15)$$

6.3.2.2 The DC-Link Voltage Controller

The aim of the DC-link voltage controller is to keep this voltage stable and according to the setpoint E^*. The output of this outer control loop provides the q-axis reference current i_{lq}^* for the inner current control loop (see Figure 6.3). The design of this controller is subject to a number of considerations or hypotheses: (i) the capacitor of the DC link must be large enough for the DC voltage dynamics to be considered as much slower than the dynamics of the inner current control loop; (ii) in turn, the inner current control loop must be fast enough for the converter to resemble an ideal three-phase current source; and (iii) losses in the power converter can be neglected.

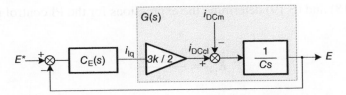

Figure 6.5 The DC-link voltage control scheme.

Considering the hypotheses formulated above, and assuming $u_{cld} = 0$ in Equation (6.4), the current entering into the DC link from the GSC can be formulated as follows:

$$i_{DCcl} = \frac{3}{2}\frac{u_{clq}i_{lq}}{E},$$ (6.16)

which, if $k = E/u_{clq}$ is considered as a constant value, results in

$$i_{DCcl} = \frac{3k}{2}i_{lq},$$ (6.17)

which depicts a linear relationship between i_{DCcl} and i_{lq}. Using this linear relationship and Equation (6.6), we can build a transfer function $G(s)$ for the DC link, which provides the DC-link voltage E, given the inputs i_{DCs} and i_{lq}. The addition of a controller $C(s)$ for the voltage E results in the closed-loop control scheme depicted in Figure 6.5.

Since i_{DCm} is normally different to zero (because it represents the current consumed or injected into the bus by an SSC), a proportional integral type controller is required in order to obtain a zero steady state error. Thus, considering a PI controller with parameters K_{pE} and K_{iE}, the closed-loop transfer function E/E^* can be formulated as follows:

$$\frac{E}{E^*} = \frac{C_E(s)G(s)}{1 + C_E(s)G(s)} = \frac{\frac{3kK_{pE}}{2C}s + \frac{3kK_{iE}}{2C}}{s^2 + \frac{3kK_{pE}}{2C}s + \frac{3kK_{iE}}{2C}},$$ (6.18)

which can be rearranged as

$$\frac{E}{E^*} = \frac{2\xi_E\omega_E s + \omega_E^2}{s^2 + 2\xi_E\omega_E s + \omega_E^2}.$$ (6.19)

Equations (6.18) and (6.19) determine the expressions for the PI control parameters, as follows:

$$K_{pE} = \frac{4C\xi_E\omega_E}{3k}$$ (6.20)

and

$$K_{iE} = \frac{2C\omega_E^2}{3k}.$$ (6.21)

As can be noted, the control parameters depend on two degrees of freedom, ξ_E and ω_E, which should be determined by the designer.

6.4 The Modeling and Control of Storage-Side Converters and Storage Containers

The previous section presented the modeling for the GSC of power conversion systems for storage technologies. As illustrated in Figure 6.1, the technology for this part of the power conversion system has been considered common to all storage installations in this chapter. Conversely, the SSCs are of different types, depending on the particular storage container under consideration. Accordingly, this section presents the modeling of various kinds of storage system, and also of the type and control method for the SSCs to which the storage containers are attached.

6.4.1 Supercapacitors and DC–DC Converters

6.4.1.1 Modeling

Supercapacitors can be modeled by the equivalent electrical circuit shown in Figure 6.6 [97, 178].

As can be noted, the supercapacitor is characterized by the capacitance C and two resistances: the series resistance, R_s, commonly called the equivalent series resistance

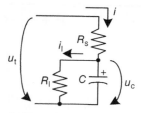

Figure 6.6   The equivalent electrical circuit of a supercapacitor.

(ESR), which is the main contributor to power loss in charging and discharging processes; and the dielectric leakage resistance, R_1. Both magnitudes can be derived from manufacturers' datasheets (for more information, see Chapter 4).

The voltage at the supercapacitor terminals can be deduced from the equivalent electrical circuit in Figure 6.6. While charging, it can be formulated as

$$u_t = u_c + i \times R_s. \tag{6.22}$$

In turn, the voltage u_c can be derived from

$$\frac{du_c}{dt} = \frac{i - i_1}{C}, \tag{6.23}$$

where i is the current drawn through the load and i_1 is the leakage current, which can be computed by

$$i_1 = \frac{u_c}{R_1}. \tag{6.24}$$

Therefore, substituting i_1 in Equation (6.23) for the expression in Equation (6.24) results in

$$\frac{du_c}{dt} = \frac{i - \frac{u_c}{R_1}}{C}. \tag{6.25}$$

While connected to the buck DC–DC converter, the topology of which was presented in Chapter 4, the coupling takes the form depicted in Figure 6.7.

If the voltage at the DC link, u_i, is considered to be constant, the system in Figure 6.7 has two states: one referring to the voltage u_c, and the other referring to the electric current through the inductor of the converter. Equation (6.25) determines the dynamics

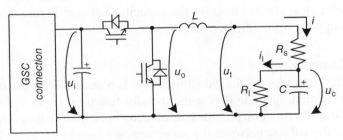

Figure 6.7 The equivalent electrical circuit of a supercapacitor connected to a DC–DC converter.

of the voltage in the capacitor. To solve the system, a second equation is needed to address the dynamics of the electric current i, and this is formulated as follows:

$$L\frac{di}{dt} + u_t = d \times u_i, \tag{6.26}$$

where $d(t)$ is the applied time-dependent duty cycle for the converter. The duty cycle varies between 0 and 1, and it relates the input and output voltages of the DC–DC converter by

$$u_o = d \times u_i. \tag{6.27}$$

The duty cycle d serves as an input to the switching scheme for the power switches of the converter; for example, a PWM switching scheme [218]. Ideally, the converter could provide any magnitude between the boundaries for u_t. However, due to practical limitations and the performance of the controller, it is preferable to constrain operation to the continuous-conduction operating region; in other words, to apply duty cycles that are high enough to ensure that the current through the inductor does not reach zero, thus entering into a discontinuous-conduction operating region. For more details on buck/boost DC–DC converters, see Wu *et al.* [219] and Kazmierkowski, Krishnam, and Blaabjerg [220].

Equations (6.25) and (6.26) can be used to build a dynamic averaged model for the coupling DC–DC converter/supercapacitors.

6.4.1.2   Control

The dynamics of the capacitor voltage (governed by Equation (6.25)) are much slower than the dynamics of the inductor current (governed by Equation (6.26)). This suggests that a cascaded controller should be designed, in which the inductor current is controlled by an inner fast control loop, which receives the current setpoints from the output of an outer slow control loop for the capacitor voltage. Figure 6.8 depicts the proposed topology for the control system.

The Current Control Loop
The design of the controller for the inductor current is based on Equation (6.26) [221]. In the Laplace domain, one can derive the transfer function, $G_i(s)$, for the DC–DC converter. This transfer function provides the inductor current, $I(s)$, function of the duty cycle, $D(s)$, and the voltage between the supercapacitor terminals, $U_t(s)$. The voltage at the DC link, U_i, is considered to be constant. The addition of a controller, $C_i(s)$, for the current $I(s)$ results in the closed-loop control scheme depicted in Figure 6.9.

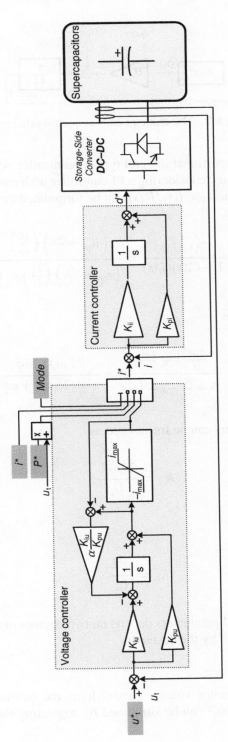

Figure 6.8 The cascaded control system for the SSC of the supercapacitor.

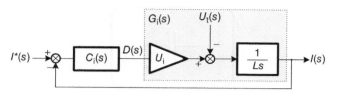

Figure 6.9 The inductor current control scheme.

Because of $U_t(s)$, a proportional integral type of controller is required to obtain a zero steady state error. Thus, considering a PI controller with parameters K_{pi} and K_{ii}, the closed-loop transfer function $I(s)/I^*(s)$ can be formulated as follows:

$$\frac{I(s)}{I^*(s)} = \frac{C_i(s)G_i(s)}{1 + C_i(s)G_i(s)} = \frac{\left(K_{pi} + \frac{K_{ii}}{s}\right)\left(\frac{U_i}{Ls}\right)}{\left(K_{pi} + \frac{K_{ii}}{s}\right)\left(\frac{U_i}{Ls}\right) + 1}, \tag{6.28}$$

which can be rearranged as

$$\frac{I(s)}{I^*(s)} = \frac{\frac{K_{pi}U_i}{L}s + \frac{K_{ii}U_i}{L}}{s^2 + \frac{K_{pi}U_i}{L}s + \frac{K_{ii}U_i}{L}} = \frac{2\xi_i\omega_i s + \omega_i^2}{s^2 + 2\xi_i\omega_i s + \omega_i^2}. \tag{6.29}$$

Thus, the control parameters can be formulated as

$$K_{pi} = \frac{2\xi_i\omega_i L}{U_i} \tag{6.30}$$

and

$$K_{ii} = \frac{\omega_i^2 L}{U_i}. \tag{6.31}$$

As can be noted, the control parameters depend on two degrees of freedom, ξ_i and ω_i, which should be determined by the designer.

The Voltage Control Loop
For the design of the capacitor voltage control loop, the equivalent circuit of the capacitor system in Figure 6.7 can be simplified by neglecting the ESR, R_s, and the

leakage resistance R_1. This is because the voltage dynamics will be mainly governed by the capacitance C. By doing this, the following relation can be established:

$$C\frac{du_t}{dt} = i, \qquad (6.32)$$

while in the Laplace domain, one can formulate the transfer function G_{uc}, which provides the supercapacitor voltage, U_t, function of the current I. Thus,

$$G_{uc}(s) = \frac{U_t(s)}{I(s)} = \frac{1}{Cs}. \qquad (6.33)$$

According to $G_{uc}(s)$, the supercapacitor, or the plant, can be considered as a pure integrator linear system, and it is proposed that this should be controlled by a PI structure with transfer function $C_{uc}(s)$. This controller will provide the reference current for the inner current control loop, as a function of the error between the desired, $U_t^*(s)$, and the actual capacitor voltages. Together, this yields the closed-loop transfer function $M(s)$, which takes the following form:

$$M(s) = \frac{U_t(s)}{U_t^*(s)} = \frac{C_{uc}(s)G_{uc}(s)}{1 + C_{uc}(s)G_{uc}(s)} = \frac{\left(K_{pu} + \frac{K_{iu}}{s}\right)\left(\frac{1}{Cs}\right)}{\left(K_{pu} + \frac{K_{iu}}{s}\right)\left(\frac{1}{Cs}\right) + 1}. \qquad (6.34)$$

Rearranging terms, the above equation can be made to resemble a second-order transfer function:

$$M(s) = \frac{\frac{K_{pu}}{C}s + \frac{K_{iu}}{C}}{s^2 + \frac{K_{pu}}{C}s + \frac{K_{iu}}{C}} = \frac{2\xi_u\omega_u s + \omega_u^2}{s^2 + 2\xi_u\omega_u s + \omega_u^2}. \qquad (6.35)$$

Now, the control parameters can be formulated by

$$K_{pu} = 2\xi_u\omega_u C \qquad (6.36)$$

and

$$K_{iu} = \omega_u^2 C. \qquad (6.37)$$

It is important to note that the voltage control system is equipped with an anti-windup (see Figure 6.8) so as to reduce the output in the integrator of the PI controller, in circumstances in which the output of this controller – that is, the reference current – cannot be produced. This could happen in the event that the reference current needs to be limited to prevent the power electronics and the supercapacitor cells from overcoming their current ratings.

6.4.2 Secondary Batteries and DC–DC Converters

6.4.2.1 Modeling

There are several approaches for modeling batteries. Among them, electrochemical, analytical, stochastic, and electrical circuit models [222] can be found in the literature. Each type represents, to a greater or lesser extent, specific phenomena in the battery cell; for example, the state of charge (SoC) and capacity variations, the temperature dependency, aging effects, and so on. However, none of these approaches is currently accurate enough to represent all of the factors affecting battery performance on its own. Thus, the modeling should be adapted to the specificities and level of detail needed for each case being studied.

Electrochemical models are based on complex nonlinear differential equations that allow us to reproduce the chemical processes within battery cells. These models are the most detailed ones amongst those considered, but also are hardly implementable in dynamic simulation environments.

Stochastic models aim to reproduce cell capacity nonlinearities from probabilistic methods based on the physical characteristics of battery cells. These models rely on Markov chains, as in Jongerden and Haverkort [224]. Despite the accuracy of these models, they are hardly implementable in dynamic simulation environments, as for the case of electrochemical models.

Analytical models aim to reproduce cell capacity nonlinearities such as the capacity rate dependency and capacity recovery phenomena using fewer equations and with a lower order than those formulated for electrochemical models. This allows us to successfully include these models in dynamic simulation environments in order to predict the SoC. However, they cannot be used to represent I–V characteristics for batteries or their dynamic behavior. Examples of analytical models are Peukert's law [225], the diffusion model [226, 227], and the Kinetic Battery Model (KiBaM) [228].

Finally, models based on electrical circuits allow us to reproduce I–V characteristics for batteries, and also their dynamic responses. To do so, these models are based on controllable voltage and current sources, in combination with (usually) variable resistances and capacitors. The limitation of these models is that they cannot estimate the SoC of the cells, so they are usually employed in conjunction with analytical models [229]. There are several approaches in the literature for models based on electrical circuits. Williamson, Rimmalapudi, and Emadi [230] summarize some of them. According to this work, the simplest approach could be that of modeling battery cells as ideal voltage sources, the open-circuit voltages of which can be obtained from field tests [231].

One step forward is to include the internal resistance of the cell in the equivalent electrical circuit [232, 233]. In this model, the internal resistance may or may not vary with the SoC and the electrolyte concentration. In any case, it is important to note that this model does not serve to represent transient states for the system (the equivalent electrical circuit can be formulated just through algebraic equations). Despite the relative simplicity of this model, it represents one of the most frequently adopted approaches for modeling battery cells. In this regard, the most representative example is the Shepherd model [234]. Because of its wide implementation and maturity, this model will be explained later in the chapter. Various updates for this model can be found in the literature. A simplified version was presented in Nasar and Unnewehr [235], and an important improvement of the model's behavior under variable load was presented in Tremblay and Dessaint [236].

Adding further complexity to the equivalent electrical circuit for the battery cell, the so-called Thevenin model has been introduced [237]. This model is composed of a voltage source in series with the internal resistance of the cell, but also includes a third element, composed of a parallel-connected capacitance with a resistance. This element represents the equivalent capacity provided by the parallel plates of the battery cell, and the corresponding resistance between the electrolyte and the aforementioned plates. All parameters characterizing the components of this scheme may or may not be considered constant. By adopting varying parameters (the function, for instance, of the SoC), several nonlinearities that occur in the battery cell can be represented.

With the aim of improving the accuracy of the model, the Thevenin equivalent circuit can be modified by adding additional lumped capacitive-resistive elements [238]. The resulting generalized equivalent circuit of order n allows us to represent specific phenomena in the cell with varying degrees of success.

Although Thevenin equivalent circuits are mainly based on lumped RC circuits, the equivalent circuit for the battery cells can present a different topology, as deduced from experimentation. For instance, the dynamic response of battery cells (of differing technologies) can be represented by an equivalent impedance that can be determined through the so-called equivalent impedance spectroscopy technique [239, 240]. In this case, the resulting equivalent impedance may resemble the Randles impedance [241] or similar.

It is important to note that since capacitors are included in the Thevenin model and in impedance-based models, these are able to represent transient states of the battery cell (e.g., the rate of change in the voltage over time under sudden load variations for the battery).

The problem with adding complexity to the model is to successfully determine its parameters for all states of the system. Einhorn *et al.* [242] propose an optimization algorithm that, using empirical data, determines the model parameters of a Li-ion battery cell function of the SoC. In particular, several current steps are applied to the real battery cell and the battery cell model. The difference between the real and simulated terminal voltages serves to iteratively run the optimization algorithm for model parametrization. The method is applied to three models: the above-mentioned model based on a voltage source in series with a resistance, and the Thevenin and second-order Randles models, depicting an increasing amount of realism as the model complexity increases. As a further example, the work performed in Dai *et al.* [243] is taken in the same direction.

As noted before, another technique for model parametrization is the so-called electrochemical impedance spectroscopy technique [239, 240, 244, 245]. Two modes are intended for deploying this technique: the so-called galvanostatic and potentio-static modes. In the former, the discrete Fourier transfer function for the equivalent impedance of the cell is deduced from evaluation of the consequences for the AC cell voltage of applying a small AC current through it. If this is done at different operating points of the cell, nonlinearities for the equivalent impedance with the temperature and the SoC can be identified. Conversely, in the latter mode, what is imposed for experimental purposes is not the current flowing through the cell but the voltage at its terminals.

Finally, a third (and simple) method to predict the model parameters is that based on the manufacturers' datasheets. To be certain, extracting the parameters in this way will serve to build up relatively simple models (such as the Shepherd model), but this is a straightforward and widely adopted strategy.

Figure 6.10 graphically summarizes the classification of battery models presented so far.

The rest of this section briefly presents the formulation of the Shepherd model (and the update performed by Tremblay and Dessaint [236]) in combination with Peukert's law [225], with the aim of providing a reference case or basic background on battery modeling.

The Shepherd Model and Peukert's Law

Because of their simplicity and maturity (this model was formulated in 1963), one of the most utilized alternatives to build up an averaged dynamic model is the Shepherd model [234]. The battery cell (or module) adopted in this model resembles a voltage source connected to a series resistance. The magnitude of the voltage source is variable and depends on the SoC and other nonlinearities. The terminal voltage of the battery while discharging is computed by the following equation:

$$
u_t = U_s - \underbrace{K\left(\frac{Q}{Q - it}\right)i}_{(i)} - \underbrace{R \times i}_{(ii)} + \underbrace{Ae^{-\frac{Bxit}{Q}}}_{(iii)}, \tag{6.38}
$$

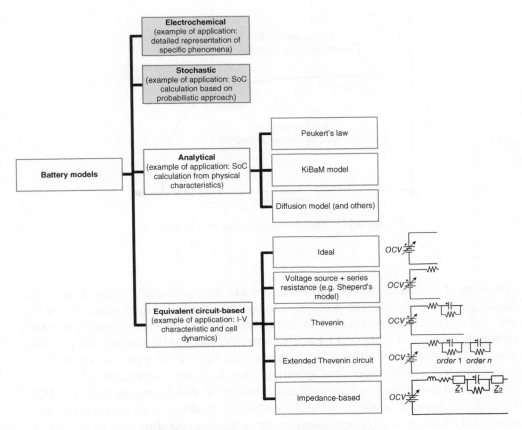

Figure 6.10   A summary of battery models.

and for charging processes

$$u_t = U_s - K\left(\frac{Q}{Q-it}\right)i - R \times i - Ae^{-\frac{B \times it}{Q}}, \qquad (6.39)$$

where u_t is the terminal voltage (in volts), i is the electric current (in amperes), Q is the maximum amount of charge for the battery (in Ah), it is the actual battery charge (in Ah), and U_s, K, R, A, and B are Shepherd parameters. These Shepherd parameters are labeled as follows: U_s is the constant voltage (in volts); K is the slope of the polarization curve, or the polarization resistance (in V/Ah); R is the internal resistance of the cell measured in the charged state; and A and B are empirical parameters.

In the Shepherd model, the electrical potential drop in the battery cell during the discharging process is divided into three parts, each of which mainly depends on

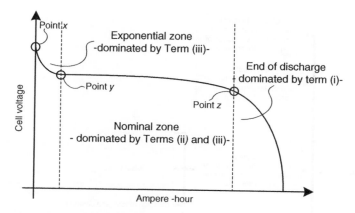

Figure 6.11   The discharge curve of a battery cell.

one term of Equation (6.38). Term (iii) depicts the nonlinear voltage drop at the beginning of the discharging process, starting from the full charge stage. Terms (ii) and (iii) determine the potential drop due to the internal resistance within the nominal operating zone (see Figure 6.11). Finally, term (i) dominates the exponential potential voltage drop while exhausting the cell charge; that is, while *it* approximates to Q.

As depicted, within the nominal operating zone, the voltage at the terminals is almost constant and this permits a relatively easy connection and management by the PCS. Finally, the dramatic voltage drop at the end of the discharge zone prevents the operation of the battery in this range, limiting the minimum operating SoC of the device.

The assumptions of the Shepherd model are as follows: (1) both the cathode and the anode have active porous materials; (2) both the electrodes and the electrolyte are homogeneous; (3) the electrodes are parallel and the current density is uniformly distributed through them; (4) the temperature is constant, as well as the internal resistance; (5) the polarization is linear over the range of discharge current densities; and (6) the discharge current is constant. This last assumption greatly affects the applicability of the model.

The model was modified by Tremblay and Dessaint [236] to extend its validity under variable charging and discharging currents. This work modified Equations (6.38) and (6.39) as follows:

$$u_t = U_s - K\left(\frac{Q}{Q-it}\right)i - \underbrace{K\left(\frac{Q}{Q-it}\right)it}_{(v)} - R \times i + \underbrace{\varepsilon}_{(iv)}, \qquad (6.40)$$

and for charging processes

$$u_t = U_s - K \left(\frac{Q}{it - \alpha \times Q} \right) i - K \underbrace{\left(\frac{Q}{Q - it} \right) it}_{(v)} - \underbrace{R \times i}_{} + \underbrace{\varepsilon}_{(iv)}, \qquad (6.41)$$

where the new term (v) is labeled as the polarization voltage. The addition of this term allows us to correct the open-circuit voltage, which is a function of the battery's SoC and this has a great impact in the voltage cell under variable discharging and charging currents. Finally, the exponential zone voltage in the Shepherd model is noted here by the term $\varepsilon(t)$, which responds to the following nonlinear system:

$$\frac{d\varepsilon}{dt} = B \times |i| \times (-\varepsilon + A \times u), \qquad (6.42)$$

where u is a binary variable that takes the value 1 for the charging mode and 0 otherwise. However, this expression is only valid for lead–acid and nickel–cadmium (Ni–Cd) batteries. For lithium-ion (Li-ion) types, it takes the form

$$\varepsilon = Ae^{-B \times it}. \qquad (6.43)$$

The model can be further modified from the Shepherd model by including the parameter α in the equation for the charging processes. According to Tremblay and Dessaint [236], $\alpha = 0.1$ seems reasonable for all of the battery types considered; that is, lead–acid, Li-ion, Ni–Cd, and nickel – metal hydride types. This value is obtained empirically.

Finally, note that for Ni–Cd and Li-ion batteries, the polarization resistance term (i) in the charge Equation (6.39) is further modified by computing the absolute value of it, and so becomes $|it|$, instead of the actual value it, to take specific phenomena into account while overcharged.

One important advantage of the Shepherd model and the adaptation by Tremblay and Dessaint is the fact that the model parameters can be derived from manufacturers' datasheets. In particular, the Shepherd parameters U_s, A, and K can be solved by evaluating the discharge Equation (6.38) at the points x, y, and z delimiting the full charge state, the exponential voltage zone, and the nominal voltage zone. These points are fully characterized in the manufacturers' datasheets. Doing this yields a system with three equations and three variables that can be solved analytically. More details on this procedure can be found in Tremblay and Dessaint [236].

As previously shown, the Shepherd model [234] serves to compute the I–V characteristics for the battery cell under steady state conditions; that is, under a constant charge and/or discharge current. To extend the applicability of the model under variable current conditions, the SoC should be properly estimated. Such an estimation can be derived by using Peukert's law in combination with the Shepherd model. According to Peukert's law, the cell capacity and the current rate are interrelated, so the higher the current rate, the lower is the capacity of the cell, and this can be clearly observed in cell datasheets. The law translates this dependency by stating that the product of the discharge current i raised to the power of n and the current discharge time constant τ_i is constant, so that

$$i^n \times \tau_i = \text{constant}. \tag{6.44}$$

The parameter n depends on the battery type and varies during the battery's life span, so it should be determined experimentally. Using this relationship, the capacity of the battery cell, Q_1, can be adjusted as a function of the actual discharge current i_1, as follows:

$$Q_1 = Q_2 \left(\frac{i_2}{i_1} \right)^{n-1}, \tag{6.45}$$

where Q_2 is the capacity (in Ah) of the cell while steadily discharged at a current rate i_2 (in A).

The adjustment of the cell capacity to current rate conditions allows us to better determine the SoC during the operation of the battery. However, the above relationship can be applied under constant current rate conditions. For application under variable operating conditions, time-dependent current profiles could be discretized so as to consider the electric current constant within each considered time step. Doing this, the SoC of the cell at the end of time interval k can be computed using

$$\text{SoC}_k = \text{SoC}_{k-1} + \Delta \text{SoC}_k, \tag{6.46}$$

and the net discharged ampere-hours result as

$$it_k = it_{k-1} + \Delta it_k. \tag{6.47}$$

The two previous equations can be applied indifferently for both charging and discharging processes. However, the computational expressions for the terms ΔSoC_k and Δit_k need to be particularized for each case and this is discussed in what follows.

The cell voltage – ampere-hour discharge curves offered in datasheets express the usable energy that can be extracted from the cell, and not the amount of energy actually stored in it. For discharging processes, for which the adjustment of the cell capacity is determined by Equation (6.45) – and this, in turn, is derived from experimental discharge curves that take just usable energy into account – there is no need to consider any charging efficiency; this is already implicit in the calculation. Thus, ΔSoC_k and Δit_k result as follows:

$$\Delta SoC_k = -\frac{i_k \times \Delta t}{3600 \times Q_2} \left(\frac{i_k}{i_2}\right)^{n-1} \tag{6.48}$$

and

$$\Delta it_k = \Delta SoC_k \times Q_k. \tag{6.49}$$

Since Equation (6.45) is not applicable to charging processes, we somehow need to consider the charge efficiency. Because of the difficulty of measuring the charge and discharge energy efficiencies for battery cells separately, manufacturers usually specify just an equivalent round-trip efficiency at the maximum depth of discharge. If this is considered as a good approximation for charging processes, Δit_k can be calculated using

$$\Delta it_k = \frac{i_k \times \Delta t}{3600} \frac{\mu}{100}, \tag{6.50}$$

where μ is the round-trip efficiency. From Δit_k, ΔSoC_k results as follows:

$$\Delta SoC_k = \frac{\Delta it_k}{it_{k-1}}(1 - SoC_{k-1}). \tag{6.51}$$

For more details on Peukert's law and its application, see Peukert [225] and Bumby, Clarke, and Forster [229].

6.4.2.2 Control

As in the case of supercapacitors and DC–DC converters, a cascaded control system based on PI controllers can be proposed for batteries. In such a structure, the inner control loop would be in charge of controlling the currents exchanged by the battery. The control loop topology and tuning are identical to those for supercapacitors, since they solely depend on the inductor included in the DC–DC converter and the voltage at the DC link that couples the SSC to the GSC. For further details of this controller, see Section 6.4.1.2.

Conversely, the outer control loop is in charge of providing the inner one with the reference profile of the current. The reference current is limited by the battery management system (BMS) for protection purposes. The reference profile of a current, and the design of the controller, greatly depend on the final application of the battery. Current profiles are further differentiated for charging and discharging processes.

Since they are somewhat related to the proper management of battery chargers, battery datasheets usually include voltage and current profiles to be applied at the

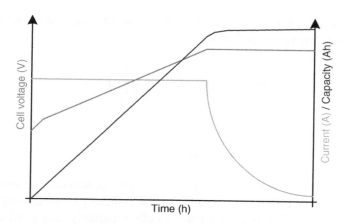

Figure 6.12 The CC/CV charging method.

battery terminals for carrying out the so-called constant-current/constant-voltage (CC/CV) charging. This charging method is widely applied in multiple applications and as such is considered worth including here as reference or background knowledge. Figure 6.12 depicts the voltage, current, and battery capacity while performing such a charging method.

As noted, in constant-current charging mode, a constant current is applied at the battery terminals – this is ensured by the current control loop of the DC–DC converter to which the battery is attached. The battery cell should be properly monitored so as to avoid overcharging and overdischarging. Once the cell voltage reaches the peak charge voltage, the controller switches to the constant-voltage mode. In this control mode, the controller aims to prevent the cell voltage from dropping from the level attained. While in this control mode, the battery cell becomes completely charged and thus the current entering in the cell drops progressively, down to a minimum value. Charge cutoff occurs as soon as predetermined minimum current is reached, indicating the full charge state of the cell. In practice, it is important to prevent the cell from overvoltage, and for this reason the controller may switch from constant current to constant voltage slightly prior to reaching the cell peak voltage.

As previously presented, the charging and discharging of the battery cell can be governed by a cascaded control system based on PI controllers. The scheme of such a structure is presented in Figure 6.13.

The tuning of the inner control loop was presented in Section 6.4.1.2. The following aims to propose an easy methodology to tune the voltage controller, and hence the outer control loop.

As presented in Section 6.4.2.1, the voltage at the cell terminals is affected by several nonlinearities within the cell, such as the dependence of the capacity on the rate of change of the current, the temperature, capacity recovery and fade, parameter variation with the SoC, and so on. For the design of voltage controller tuning, it would be desirable to characterize the system or the plant to control – that is, the battery

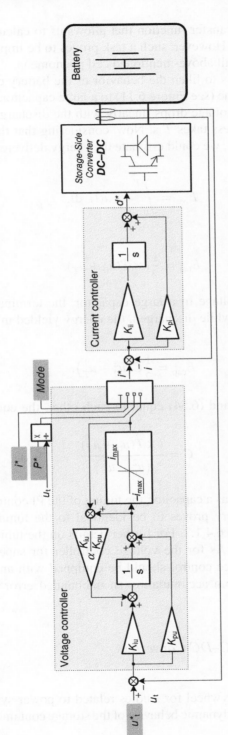

Figure 6.13 The control scheme for the SSC attached to the battery.

cell – by an equivalent transfer function that allows us to calculate the cell voltage from the electric current. However, such a task proves to be impracticable due to the difficulty of considering all above-mentioned cell phenomena.

One simple approach is to liken the behavior of the battery cell voltage while in the nominal operating zone (see Figure 6.11) to a large capacitor. Within the nominal operating zone, the cell voltage drops linearly with the discharged capacity from u_0 down to u_T. Such a process takes T s. Now, considering that the cell is discharged steadily at a given current, we could estimate the energy delivered, E_{bat} (in watts), by the cell as

$$E_{bat} = \int_0^T i \times u(t) \ dt, \tag{6.52}$$

so that

$$E_{bat} = \frac{1}{2} i \times T(u_0 + u_T). \tag{6.53}$$

Now likening the cell voltage to a large capacitor, the terminal voltage of which also drops from u_0 to u_T while discharged, the energy yielded in this process can be computed as

$$E_{cap} = \frac{1}{2} C \left(u_0^2 - u_T^2 \right). \tag{6.54}$$

Setting Equations (6.53) and (6.54) equal to each other, the equivalent capacity C becomes

$$C = \frac{i \times T(u_0 + u_T)}{\left(u_0^2 - u_T^2 \right)}. \tag{6.55}$$

Considering the battery as a capacitor, the tuning of the PI controller of the voltage control loop for the battery proves to be identical to the tuning of that loop for supercapacitors in Section 6.4.1.2. For further details on the tuning process, see the above-mentioned section. As for the voltage controller for supercapacitors, the PI structure for battery voltage control should be equipped with an anti-windup. This prevents the integrator from accumulating an unbounded error while in constant-current control mode.

6.4.3 Flywheels and AC–DC Converters

6.4.3.1 Modeling

An averaged model of a flywheel for studies related to power systems can serve to represent the averaged and dynamic behavior of the storage container while interacting with the electrical network. Thus, such an averaged model should present electrical magnitudes (voltages and currents) as input–output signals, so as to effectively enable the integration of the model into the relevant electrical simulation platforms.

Although the topology of a flywheel-based storage system has previously been presented in Chapter 4, it is convenient to recall here that it is composed of a rotating disk, coupled to an electrical machine, which in turn is connected to the electrical grid through a set of back-to-back power converters. In what follows, the modeling of the mechanical part of the system and the electrical modeling of the rotating machine and the storage-side power converter are introduced.

Since a flywheel stores electrical energy as mechanical energy, it is mainly characterized by two quite different time constants: the mechanical time constant, determined by the inertia of the system; and the electrical time constant, determined by the electric and electronic parts of the technology. By modeling the mechanics of the system, one can reliably represent the major parameters of the storage container, such as the SoC – which is directly related to the speed of the flywheel – and the mechanical torque at the shaft of the electrical machine to which the rotating disk is coupled. Furthermore, by making some assumptions about the electrical efficiencies of the system, one could derive the electric power actually exchanged with the system to which the storage is connected; that is, the main grid.

The large time constant of the mechanics, governed by the inertia of the rotating parts, mainly drives the dynamics of the system. The equation of motion of the electromechanical system comprised of the electrical machine and the rotating disk can be compared to that of a single-mass system as follows:

$$\frac{d\omega_r}{dt} = \frac{p}{2} \frac{T_e - T_1}{J},\tag{6.56}$$

where ω_r is the electrical frequency of the machine, in rad/s, p is the number of poles, T_e is the electrical torque produced, in Nm, T_1 is the mechanical torque at the shaft of the machine, and J is the inertia of the rotating parts, in kgm^2.

The mechanical torque T_1 can be associated with frictional losses in the assembly and is speed dependent, so

$$T_1 = k_1 \times \omega_m + k_2,\tag{6.57}$$

where k_1 and k_2 are parameters to be adjusted empirically and/or from manufacturers' datasheets. The mechanical speed ω_m, in rad/s, can be expressed in terms of the electrical frequency and the number of poles by

$$\omega_m = \omega_r \frac{p}{2}.\tag{6.58}$$

The general equation for computing the electrical torque T_e, which is valid for both surface-mounted (or nonsalient pole) and salient pole machines, is

$$T_e = \left(\frac{3}{2}\right)\left(\frac{p}{2}\right)(\psi_{PM}i_{sq} + (L_d - L_q)i_{sq}i_{sd}),\tag{6.59}$$

where i_{sqd} are the stator currents, in amperes; and L_{qd} are the stator inductances, in henries, which depend on leakage and magnetizing inductances, so that $L_q = L_{ls} + L_{mq}$ and $L_d = L_{ls} + L_{md}$; and ψ_{PM} is the flux generated by the permanent magnets, in webers (Wb).

Since a surface-mounted permanent magnet machine is being considered in this case, the direct-axis and quadrature-axis inductances of the machine are approximately equal: $L_d \approx L_q$. Accordingly, the electrical torque can be expressed as follows:

$$T_e = \left(\frac{3}{2}\right)\left(\frac{p}{2}\right)\psi_{PM}i_{sq}.\tag{6.60}$$

The electric currents, magnetic fluxes, and inductances are associated with the modeling of the electrical part of the system. This electrical model mainly comprises the voltage equations for a permanent magnet synchronous machine (PMSM) – this is one of the most widely utilized drivers for flywheels.

The voltage equations are usually introduced in Park's $qd0$ coordinates [213], instead of using alternative abc magnitudes. In the rotor's frame of reference, with stator currents in motor orientation, and expressing all magnitudes in SI units, the voltage equations are formulated as follows:

$$u_{sq} = R_s i_{sq} + L_q\frac{d}{dt}i_{sq} + \omega_r L_d i_{sd} + \omega_r \psi_{PM},\tag{6.61}$$

$$u_{sd} = R_s i_{sd} + L_d\frac{d}{dt}i_{sd} - \omega_r L_q i_{sq},\tag{6.62}$$

$$u_{0s} = R_s i_{0s} + L_{ls}\frac{d}{dt}i_{s0},\tag{6.63}$$

where the u_{sqd0} represent stator voltages, in volts, and R_s is the stator resistance, in ohms.

The stator terminals of the PMSM are connected to an inductive filter, to filter the harmonic components of the currents flowing into the servomotor. This inductive filter, in turn, couples the servomotor to the SSC. Considering the converter as three ideal voltage sources (u_{csa}, u_{csb}, and u_{csc}), the equivalent electric circuit of the machine-side converter can be made to resemble that presented in Figure 6.14.

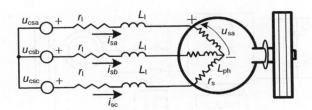

Figure 6.14 The machine-side converter circuit.

In a synchronous $qd0$ frame of reference with the electrical frequency of the servomotor, the mathematical modeling of the system presented in Figure 6.14 takes the following form:

$$u_{\text{csq}} - u_{\text{sq}} = r_1 i_{\text{sq}} + L_1 \frac{\text{d}}{\text{d}t} i_{\text{sq}} + \omega_\text{r} L_1 i_{\text{sd}}, \tag{6.64}$$

$$u_{\text{csd}} - u_{\text{sd}} = r_1 i_{\text{sd}} + L_1 \frac{\text{d}}{\text{d}t} i_{\text{sd}} - \omega_\text{r} L_1 i_{\text{sq}}. \tag{6.65}$$

Substitution of Equations (6.61)–(6.62) in Equations (6.64)–(6.65) results in

$$u_{\text{csq}} = (r_1 + r_\text{s}) i_{\text{sq}} + (L_1 + L_\text{q}) \frac{\text{d}}{\text{d}t} i_{\text{sq}} + \omega_\text{r} (L_1 + L_\text{d}) i_{\text{sd}} + \omega_\text{r} \psi_{\text{PM}}, \tag{6.66}$$

$$u_{\text{csd}} = (r_1 + r_\text{s}) i_{\text{sd}} + (L_1 + L_\text{d}) \frac{\text{d}}{\text{d}t} i_{\text{sd}} - \omega_\text{r} (L_1 + L_\text{q}) i_{\text{sq}}. \tag{6.67}$$

The set of equations presented above builds up the mechanical and electrical modeling of the system.

6.4.3.2 Control

This section presents the design of the control system for the SSC of a flywheel-based storage system. As previously introduced, this controller is in charge of managing the speed of the flywheel, and thus, the energy injected or absorbed by the storage container from the main grid. Also, the controller manages the reactive currents flowing between the stator and the SSC terminals.

The control scheme adopted for these purposes is the so-called current vector control algorithm for the PMSM [213, 246, 247], which is graphically presented in Figure 6.15.

As can be observed, as for the GSC, the controller presents a cascaded architecture, in which the outer loop contains the speed controller, feeding the q-axis inner current control loop. The input measurements are the alternating currents flowing from the converter, i_{sabc}, the rotor angle, θ_r, and the DC-link voltage, E. The control setpoints

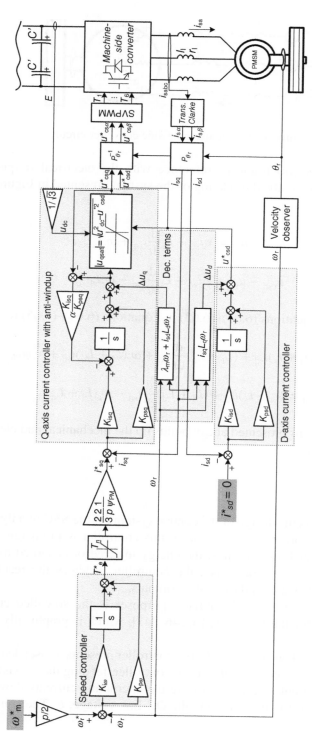

Figure 6.15   The control scheme for the machine-side converter.

are the mechanical speed of the PMSM, ω_m^*, and the d-axis stator current component, i_{sd}^*. The outputs of the algorithm are the voltage setpoints for the SVPWM scheme for the converter [213], expressed in a stationary frame of reference as $u_{cs\alpha\beta}^*$.

As indicated in Equation (6.60), the electromagnetic torque produced by the machine for acceleration and breaking is only dependent on i_{sq}, so this justifies the association of the speed controller with, precisely, the determination of the q-axis stator current setpoint, i_{sq}^*.

On the other hand, the d-axis stator currents do not impact on the speed control, provided that the machine does not overcome its rating speed. Thus, normally, $i_{sd}^* = 0$ (no reactive currents are exchanged through the inductive filter between the SSC and the stator terminals). However, with the aim of overcoming the speed ratings of the machine, nonzero i_{sd}^* can be applied, thus adopting the concept of flux weakening [248, 249].

The following paragraphs address the design of the PI controllers, building up both the outer and the inner control loops in Figure 6.15.

The Current Control Loops

As presented in Equations (6.66) and (6.67), the q-axis component of the stator voltage u_{sq} depends on the d-axis component of i_{sd}. Analogously, the u_{sd} depends on i_{sq}. In order to obtain a decoupled linear system, the following restructuring is proposed:

$$u_{csq} = \hat{u}_{csq} + \underbrace{\omega_r(L_1 + L_d)i_{sd} + \omega_r\psi_{PM}}_{\Delta u_q}, \tag{6.68}$$

and

$$u_{csd} = \hat{u}_{csd} - \underbrace{\omega_r(L_1 + L_q)i_{sq}}_{\Delta u_d}, \tag{6.69}$$

where $\hat{u}_{csq}$ and $\hat{u}_{csd}$ describe a decoupled linear system, $G(s)$, which takes the following form in the Laplace domain:

$$i_{sqd} = \underbrace{\begin{pmatrix} \frac{1}{(r_1+r_s)+(L_1+L_q)s} & 0 \\ 0 & \frac{1}{(r_1+r_s)+(L_1+L_d)s} \end{pmatrix}}_{G(s)} \hat{u}_{csqd}. \tag{6.70}$$

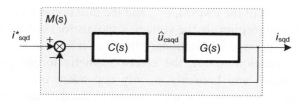

Figure 6.16 The block diagram control methodology for the stator currents.

Since $G(s)$ describes a linear system, a PI controller $C(s)$ is proposed to control the stator currents of PMSM, yielding the closed-loop structure block diagram shown in Figure 6.16.

As for the tuning of the PI controllers for the GSC, the direct synthesis method [217] is used for tuning these stator current controllers. As a reminder, this method consists of determining the controller $C(s)$, the transfer function of which is specified according to the desired closed-loop transfer function $M(s)$. With regard to the desired performance, $M(s)$ is preferred to be in the form of a first-order mathematical function dependent on a parameter λ_s. Proceeding as in Section 6.3.2.1, the proportional and integral control parameters for current control PI structures are as follows:

$$K_{psq} = \frac{(L_1 + L_q)}{\lambda_s} \tag{6.71}$$

and

$$K_{isq} = \frac{(r_1 + r_s)}{\lambda_s}. \tag{6.72}$$

As presented in Figure 6.15, the output of the PI controllers, $\hat{u}_{csd}$, should be complemented by the decoupling terms Δu_q and Δu_d defined above, to determine the control voltages u_{csq} and u_{csd} for the convertor's SVPWM control scheme. In addition, the q-axis PI structure is also equipped with an anti-windup. This is added so as to avoid instabilities in the SSC, which are eventually induced by the variable DC-link voltages. To understand such instabilities, it is important to first note some of the limitations of the SVPWM scheme.

In an SVPWM scheme, the modulus of the peak AC phase-neutral voltage, $u_{ref}|$, that can be produced (within the normal linear modulation range) is dependent on the magnitude of the voltage at the DC link. This linear dependency is formulated by $|u_{ref}| \leq E/\sqrt{3}$. In turn, $|u_{ref}|$, which can be made to resemble the magnitude of the stator voltage of the PMSM, depends linearly on the speed of the machine. Thus, the maximum speed of the PMSM could be bounded by the DC-link voltage E, provided that this becomes reduced under certain circumstances.

In other words, in the event of a deep drop in the DC-link voltage, the SVPWM scheme would not be able to develop a high enough sinusoidal voltage at the stator terminals to maintain the desired speed, thus causing the integrator of the PI controller to fall into an oversaturated state. Once the DC-link voltage recovers, and due to the oversaturated state of the controller, the servomotor will accelerate uncontrollably. The addition of the anti-windup serves to limit the output of the integrator, thus avoiding potential instabilities.

In practice, and as depicted in Figure 6.15, for the implementation of the anti-windup scheme, the q-axis reference voltage u^*_{csq} determined by the PI controller is a saturated function of E and u^*_{csd}, of the form

$$|u^{sat}_{csq}| = \sqrt{\left(\frac{E}{\sqrt{3}}\right)^2 - (u^*_{csd})^2}. \tag{6.73}$$

In the event that the q-axis PI controller determines a reference voltage u^*_{csq} higher than its threshold magnitude u^{sat}_{csq} – that is, in the event that a rotational speed for the machine being demanded that is higher than that permitted according to the DC bus voltage – the input for the integrator of the controller is diminished so as to avoid this developing into an oversaturated state. The diminishing term e is proportional to the difference between the voltage u^*_{csq} that the PI controller aims to achieve and the maximum voltage that the converter determines as admissible, all affected by a factor α, which depends on the control parameters K_{psq} and K_{isq}. In particular, the signal e is computed as follows:

$$e = (u^*_{csq} - u^{sat}_{csq})\alpha, \tag{6.74}$$

where $\alpha = K_{psq}/K_{isq}$.

Speed Control

To complete the description of the SSC for a flywheel-based power conversion system, this section introduces the design of the outer control loop for the above-mentioned converter. This controller is in charge of regulating the speed of the flywheel (see Figure 6.15).

The speed control is deployed by adjusting the torque of the PMSM. The reference torque is proportional to the q-axis stator current i_{sq}, according to Equation (6.60). An ideal orientation of the $qd0$-axis with the magnetic flux created by the magnets of the rotor of the PMSM, and a current controller bandwidth well beyond the required speed controller bandwidth, are assumed.

For controller design, the transfer function of the "plant" to be controlled mainly corresponds to the Laplace representation of the mechanical dynamics of the system

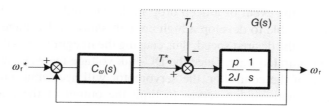

Figure 6.17 A rotational speed control scheme.

$G(s)$, which were formulated in Equation (6.56). Graphically, the closed-loop representation for the design of the speed controller is presented in Figure 6.17, in which $C_\omega(s)$ is the speed controller.

Provided that the torque losses T_l equal zero, the controller $C_\omega(s)$ could take the form of a proportional gain, affecting the error $e = \omega_r^* - \omega_r$. Such a proportional controller would cancel any steady state error in response to a step-profiled reference speed ω_r^*, since the open-loop transmittance $C_\omega(s)G(s)$ features a pure integrator:

$$C_\omega(s)G(s) = K_{p\omega}\frac{p}{2Js}. \tag{6.75}$$

However, since T_l is not negligible, the speed regulator should also consider an integrator to obtain a zero steady state error. Nevertheless, here, a proportional controller is considered. In any case, it is important to note that the output of the controller should be limited at the minimum value between the maximum admissible torque for the PMSM and the maximum admissible currents for the converter.

Considering a proportional controller, the closed-loop transfer function of ω_r/ω_r^* results in

$$\frac{\omega_r}{\omega_r^*} = \frac{C_\omega(s)G(s)}{C_\omega(s)G(s) + 1} = \frac{\frac{K_{p\omega}p}{2Js}}{\frac{K_{p\omega}p}{2Js} + 1} = \frac{1}{\frac{2J}{K_{p\omega}p}s + 1}, \tag{6.76}$$

which can be made to resemble a first-order transfer function:

$$\frac{\omega_r}{\omega_r^*} = \frac{1}{\lambda_\omega s + 1}. \tag{6.77}$$

Finally, the proportional parameter of the controller is formulated as being dependent on the value selected for the time constant λ_ω:

$$K_{p\omega} = \frac{2J}{p\lambda_\omega}. \tag{6.78}$$

6.5 An Example of an Application: Discharging Storage Installations Following Various Control Rules

6.5.1 Input Data

This section depicts a simple example of an application, so as to demonstrate the performance of the models and controllers previously presented for supercapacitors, batteries, and flywheels. The adopted input data for the modelization are presented in Tables 6.1, 6.2, and 6.3.

The parameters for the flywheel motor are listed in Díaz-González *et al.* [250].

The supercapacitor storage container is built up from multiple modules from Maxwell Technologies, Inc. [148]. Each module, with reference BMOD0058 E016

Table 6.1 Parameters for simulation (I/III).

System	Component(s)	Parameter	Value
External grid		U_1 (V)	400
		f (Hz)	50
GSC	Inductive filter	L_1 (H)	0.005
		R_1 (Ω)	0.3
	DC link	E_0 (V)	800
		C (F)	0.01
	Controllers	K_{pPLL}	1.0
		K_{iPLL}	1.0
		K_{ilqd}	300.0
		K_{plqd}	5.0
		K_{pE}	0.2956
		K_{iE}	1.7055
Flywheel	Mechanics	ω_{max}(krpm)	3000.0
		ω_{min}(krpm)	1000.0
		$T_{acc_{max}}$ (Nm)	18.0
		$T_{acc_{min}}$ (Nm)	−18.0
		J(kgm^2)	2.6145
		c_1 (Ws^2/rad^2)	0.0019
		c_2 (Ws/rad)	0.282
	PMSM	ψ_m(Wb)	0.2465
		L_d and L_q(H)	0.0029
		R_s(Ω)	0.44
		p	6
		P_{rated}(kW)	5.5
	SSC	K_{psqd}	2.9
		K_{isqd}	440.0
		$K_{p\omega}$	1.743

Table 6.2 Parameters for simulation (II/III).

System	Component(s)	Parameter	Value
Supercapacitor	Modules	C (F)	1.8125
		U_{max} (V)	512
		U_{min} (V)	250
		R_s (Ω)	0.704
		R_l (Ω)	1000
		P_{rated} (kW)	44.3
	SSC	L (H)	0.005
		K_{pi}	2.5×10^{-4}
		K_{ii}	0.0025
		K_{pu}	7.25
		K_{iu}	7.25

B02, offers 58 F of capacity and 16 V between its terminals. By connecting 32 modules in series, the system acquires the ratings presented in Table 6.2. It is important to note that for series-connected modules, the equivalent capacity results from

$$C = \frac{C_{module}}{\text{number of modules in series}} \tag{6.79}$$

and the ESR results from

$$R_s = (R_{s\_module}) \times (\text{number of modules in series}). \tag{6.80}$$

Table 6.3 Parameters for simulation (III/III).

System	Component	Parameter	Value
Battery (Li-ion)	Cell	R (Ω)	0.005
		u_{max} (V)	3.95
		u_{exp} (V)	3.85
		u_{nom} (V)	3.3
		u_{min} (V)	2.7
		Q (at 0.33 C, in Ah)	41
	SSC	L (H)	0.005
		K_{pi}	2.5×10^{-4}
		K_{ii}	0.0025
		K_{pu}	4.32×10^5
		K_{iu}	2.16×10^5

The parameters for the battery cell correspond to model VL41M from Saft [161]. The voltages u_{max}, u_{exp}, u_{nom}, and u_{min} have been derived from the discharge profile in the product datasheet. These values correspond to the voltage at full charge state, at the end of the exponential zone voltage, at the end of the nominal zone, and for the minimum admissible operating voltage, respectively. In the case of having a battery module with n_s cells in series and n_r cells in parallel, to determine the parameters for the battery module, all voltages have to be multiplied by n_s, all currents by n_r, all resistances by the factor n_s/n_r, and all capacitances by $n_s \times n_r$.

6.5.2 Discharge (Charge) Modes for Supercapacitors

This section presents the behavior of supercapacitors while discharged (charged) at constant current, at constant power, and in voltage control mode (i.e., while governed by the SoC controller).

The simulation results depict the voltage at the supercapacitor terminals, the exchanged active power, and the electric current. Figure 6.18 presents the

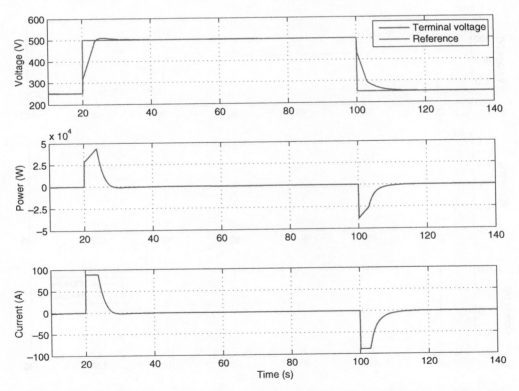

Figure 6.18 The voltage, active power, and current for the supercapacitor module while charged and discharged in voltage control mode.

behavior of the supercapacitor while discharged (charged) in voltage control mode. As a reminder, the SoC of the system is directly proportional to the voltage between the terminals of the supercapacitor module. As can be noted, the exchanged active power increases with the voltage at the supercapacitor terminals (i.e., with the SoC of the system). This relationship between the SoC and the power capability of both supercapacitors and flywheels suggests that the SoC controllers for these storage systems should be designed taking the power demand profiles they should have to satisfy into account. For instance, while building up uninterruptible power supply systems (UPSs), supercapacitors and flywheels are required to be rapidly discharged at full power capability in the event of a mains failure. So in this case, they should remain fully charged in steady state conditions. Conversely, in applications related to power smoothing, in which the storage containers are supposed to be continuously charged and discharged (see Chapter 7, which describes one example in this regard), they are better operated partially charged.

The simulation in Figure 6.19 depicts the behavior of supercapacitors while discharged (charged) in current control mode. As shown, the temporal trends for the

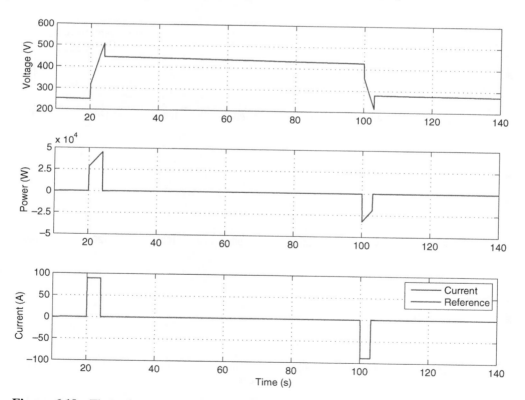

Figure 6.19 The voltage, active power, and current for the supercapacitor module while charged and discharged in current control mode.

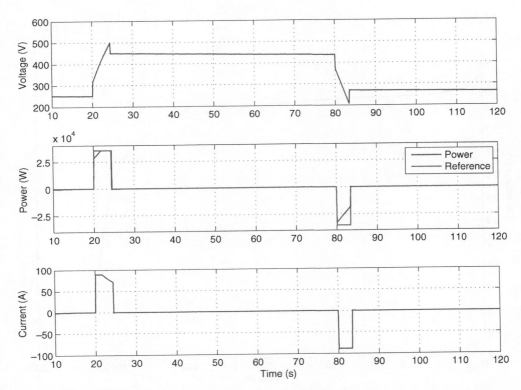

Figure 6.20 The voltage, active power, and current for the supercapacitor module while charged and discharged in power control mode.

terminal voltage, the power, and the current are quite similar to those obtained in voltage control mode. Also, note the sudden voltage drop and increase at the time when the charging and discharging processes finish, respectively, and so when the current reaches zero. These abrupt changes are due to the voltage drop at the series resistance of the supercapacitor module.

Finally, Figure 6.20 plots the behavior of the system while discharged (charged) in power control mode. Note how the power exchanged is clearly dependent on the terminal voltage of the supercapacitor module.

6.5.3 Discharge (Charge) Modes for Batteries

This section presents the behavior of a battery cell while charged in the CC/CV charging mode. The battery parameters are presented in Table 6.2. The adopted battery model is the Shepherd model as updated by Tremblay and Dessaint [236], which was briefly presented in Section 6.4.2.

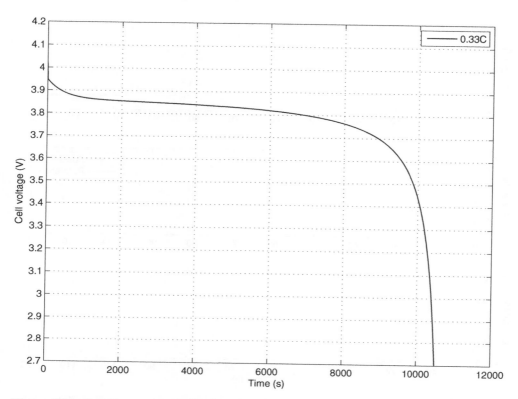

Figure 6.21 A typical voltage-discharge curve for a battery cell (the discharge rate is 0.33 C).

First, however, Figure 6.21 presents the voltage-discharge capacity for the cell while discharged in constant-current mode at 0.33 C. This discharge trend is quite similar to that presented in the manufacturer's datasheet [161] (as a reminder, the battery cell simulated corresponds to Saft's model VL41M).

The simulation results shown in Figure 6.22 depict the cell voltage, the current, and the charged capacity. The cell is charged at 0.33 C up to approximately 94% of maximum capacity.

As can be noted, the controller switches from constant-current to constant-voltage mode once the cell has reached the voltage setpoint (around 4.05 V). From this point on, the controller concentrates on preventing the cell voltage from dropping, and the current decreases progressively up to the end of the charging process.

6.5.4 Discharge (Charge) Modes for Flywheels

This section shows the behavior of flywheels while discharged (charged) at constant current (or constant torque), at constant power, and in speed control mode (i.e., while

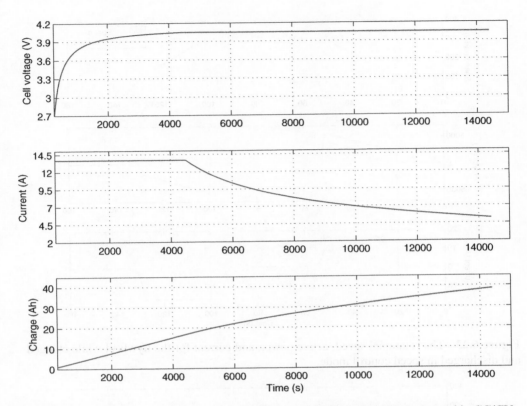

Figure 6.22 The battery voltage, current, and charge (in Ah) while being charged in CC/CV charging mode.

governed by the SoC controller). These are the typical control modes for flywheels while providing various services.

The simulation results in Figure 6.23 depict profiles for the rotational speed, the active power, and the q-axis current (which is proportional to the torque) while charged and discharged between admissible speed limits. As can be noted, and due to the large inertia of the system, the mechanical dynamics are much faster than those for the electric current of the servomotor. Also, it is interesting to note how the active power exchanged by the system increases progressively with the rotating speed.

Figure 6.24 depicts the performance of the system while discharged and charged under torque control. In this case, the q-axis electric current is always bounded by the ratings of the servomotor and the active power increases with the rotational speed of the flywheel. As shown, the obtained profiles are quite similar to those presented in Figure 6.23, corresponding to the speed control mode.

The constant-power discharging mode is noticeably different to the previous control modes. As shown in Figure 6.25, the system is able to follow the squared-profile

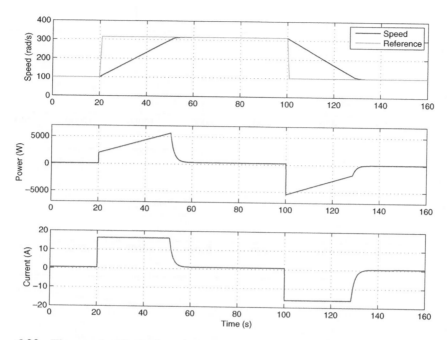

Figure 6.23   The rotational speed, active power, and current for the flywheel while charged and discharged in speed control mode.

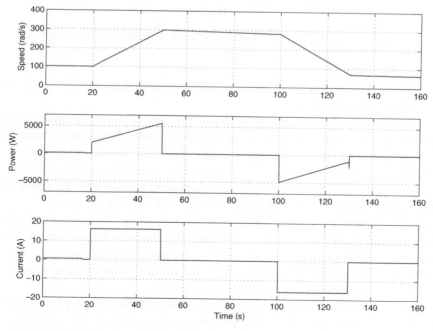

Figure 6.24   The rotational speed, active power, and current for the flywheel while charged and discharged in constant-current (or torque) control mode.

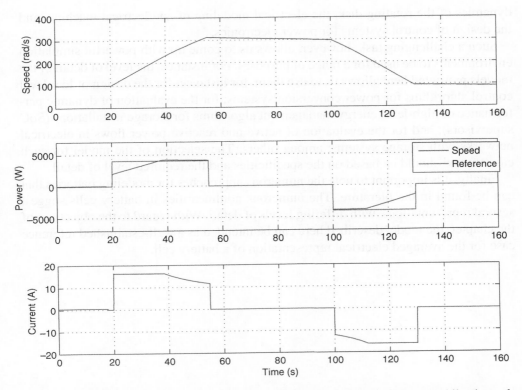

Figure 6.25 The rotational speed, active power, and current for the flywheel while charged and discharged in power control mode.

reference power provided that the rotating speed does not fall below a certain level. Under such circumstances, the power exchanged by the system decreases progressively with the rotational speed, and the torque proves to be bounded by the electrical ratings of the servomotor and the attached power electronics.

6.6 Conclusions

This chapter has presented averaged models for the dynamic simulation of various storage technologies. The modeling covers both the storage containers and also the design and tuning for the controllers of the associated power conversion systems to which they are attached. As a result, the modeling of the storage systems presented in this chapter (supercapacitors, batteries, and flywheels) has required knowledge in various fields, such as electrical, mechanical, chemical, and control engineering. For instance, the modeling of flywheels demands the representation of the mechanical dynamics of the rotating disk, the electrical modeling of an electrical machine, and the design of control systems for power electronics.

Such a challenging task, however, allows us to come up with powerful simulation environments to be used for a range of purposes. For instance, the level of detail of the models presented here allows us to use them, for instance: for the design of low-level control algorithms for power conversion systems; for the evaluation of dynamic performance of high-level energy management algorithms for storage installations (SoC supervisors); and for the evaluation of active and reactive power flows in electrical networks with storage systems; among others. The selection of the model for each case studied should be based on the specificities and the required level of detail.

Finally, it is important to note the numerous approaches for modeling batteries that can be found in the literature. The numerous nonlinearities in battery cells suggest several representations, with differing levels of detail. With regard to the objectives of the chapter, the model described here can be intended as a well-established reference case for the averaged electrical representation of a battery cell.

7

Short-Term Applications of Energy Storage Installations in the Power System

7.1 Introduction

This chapter describes potential short-term applications that energy storage systems (ESSs) could provide in the electric power system. As discussed throughout the book, the inclusion of energy storage installations in the power system is usually motivated by the need to ensure the required balance between generation and demand, as well as to satisfy the required power quality levels, regardless of the variability of renewable energy sources. Since the electric power systems of the future will surely be characterized by holding increasing penetration rates of renewables, most of the applications for storage systems discussed hereinafter will be closely related to renewable generation. Each technical issue has been identified and defined according to Barton and Infield [160], Bayod-Rújula [251], Beaudin *et al.* [166], Dell and Rand [135], EPRI [252], Georgilakis [253], and Świerzyński *et al.* [254]. In addition, the definition of these aspects is complemented by a brief discussion on the role of the ESS in each case.

The subjects presented above will be discussed in the following sections. First, however, and with the aim of providing an initial general picture, the short-term services that EESs can provide in the power system are listed and classified by storage technology in Tables 7.1 and 7.2. The literature cited in these tables mostly relates EESs with wind power (for further details, see our previous work [42]). Moreover, the tables only include short-term applications; that is, those requiring the ESSs to provide or absorb energy continuously for less than 30 min. For discussions on mid-and long-term applications, see Chapter 8.

Energy Storage in Power Systems, First Edition. Francisco Díaz-González, Andreas Sumper and Oriol Gomis-Bellmunt.
© 2016 John Wiley & Sons, Ltd. Published 2016 by John Wiley & Sons, Ltd.

7.2   A Description of Short-Term Applications

Each of the short-term applications in Tables 7.1 and 7.2 is discussed in the following sections.

7.2.1   Fluctuation Suppression

Fast output fluctuations (in the time range up to a minute) of renewable-based generating systems can cause network frequency and voltage variations, especially in isolated power systems, thus impairing the power quality [265]. In order to mitigate the effects of power fluctuations, an ESS can be used. Storage technologies suitable for this application present high ramp power rates and high cycling capability, since fast power modulation and continuous operation are required. Thus, batteries (excluding conventional lead–acid batteries), flow batteries, and especially short-timescale forms of energy storage such as supercapacitors, flywheels, and superconducting magnetic energy storage (SMES) systems are well suited for this service.

A widely accepted solution to mitigate the power fluctuations of a wind turbine driving a doubly fed induction generator (DFIG) is to include an ESS in the DC link of the back-to-back converters of the machine. This storage device is equipped with a control that interacts with the turbine's and other controls in order to optimize the net power delivered to the external grid by the entire system. This is the case presented in Qu and Qiao [287]: a supercapacitor connected to the DC link of a

Table 7.1 An overview of publications regarding the uses of ESSs in the field of wind power (part I).

	Storage duration at full power (min)	PHS	HESS	CAES	VRB	ZBB	PSB
Fluctuation suppression	≤ 1				[255–258]	[131, 255]	[255]
LVRT	≤ 1				[256, 259]	$\sqrt{a}$	$\sqrt{a}$
Voltage control support	≤ 1				[255–257, 260]	[255]	[255]
Oscillation damping	≤ 1		[261]		[255]	[255]	[255]
Spinning reserve	1–30	[91]	[262]	[263]	[255, 257, 258, 260, 264]	[131, 255, 260, 264]	[255, 260, 264]

[a]Although the storage technology is suitable for this application, dedicated studies are not listed here.

Table 7.2 An overview of publications regarding the uses of ESSs in the field of wind power (part II).

	Storage duration at full power (min)	NaS	Lead acid	Ni-Cd	Li-ion	SMES	FESS	SCESS
Fluctuation suppression	≤1	[266]	[267, 268]	[267, 268]	[162, 269]	[270–276]	[248, 277–284]	[281, 285–287]
LVRT	≤1	[156, 288, 289]	[288, 289]	[288, 289]	[288, 289]	[141, 274, 276, 285, 290–294]	[141]	[172, 285]
Voltage control support	≤1	[288, 295]	[295–298]	[267, 268, 295]	[288, 295]	[141, 270, 293, 299]	[141, 282]	[178]
Oscillation damping	≤1	[289, 300, 301, 303]	[289, 300, 303]	[289, 300, 303]	[289, 300, 303]	[293, 294, 299, 300, 302, 304]	[305, 306]	[307]
Spinning reserve	1–30	[101, 308, 309]	[296, 308–310]	[308, 309, 311]	[308, 309]	[223]	[260, 312]	

wind generator through a two-quadrant DC–DC converter. Two levels of control are defined, the high level (the wind farm supervisory controller), which is in charge of coordinating the setpoints of each wind generator, and the low level, which details the vector controllers of the converters of each wind generator. As a wind turbine controller, the power conversion system (PCS) of each storage device receives the setpoint calculated by the high-level controller, and manages the power injection or absorption by means of computing the difference between this signal and the actual active power of the wind generator.

Flywheels are also under study as a way of complementing the DC link of DFIG wind turbines. Since the operating principle of this technology is highly related to the power management of a motor/generator, the control theory of electrical drives plays a key role in these studies. In Boukettaya, Krichen, and Ouali [277], three techniques of sensorless vector-controlled induction motors driving a flywheel are compared. In addition, control theories based on the Model Reference Adaptive System (MRAS) for the design of speed estimation algorithms, as well as flux-weakening aspects, motivated by the high speed of the flywheel, are taken into account [248, 283]. In addition to the use of induction machines, permanent magnet and switched reluctance machines are being studied for flywheel storage devices [284].

Looking at new power system topologies, the combination of storage systems, such as flywheels, supercapacitors, or batteries, in hybrid systems with offshore wind generation, diesel, and photovoltaic generation is proposed by Ray, Mohanty, and Kishor [281].

Other studies [272, 273] propose the use of SMES systems in order to perform the task of fluctuation suppression, providing storage at the point of common coupling (PCC) of a wind farm to the network. In this configuration, the rated power of SMES systems reaches several MW. For instance, a 15 MWh – 60 s SMES system is proposed in Nomura *et al.* [275], in order to smooth the power fluctuations of a 100 MW wind power installation. In this case, the wind power plant (WPP) is connected to the external grid through a back-to-back DC link. To conclude, it is noted that by means of the management of the charge and discharge rates of SMES devices, the capacity of the power converters of the WPP can be reduced by 60%. Issues such as SMES capital costs, as well as power losses due to maintaining a low operating temperature and leakage of magnetic fields, have to be taken into account.

This chapter includes a particular study on the application of flywheels for wind power smoothing (see Section 7.3).

7.2.2   Low-Voltage Ride-Through (LVRT)

Voltage control of renewable-based power plants at the point of connection with the external grid during voltage dips is carried out in order to prevent the plant from being disconnected, which could cause the collapse of the network. For this reason, grid codes require wind parks to withstand voltage dips up to 0% of the rated voltage and for a specified duration. These requirements are known as LVRT requirements. Since many renewable-based generating systems are connected to the grid through power converters, it is possible to adjust the reactive power output during these disturbances [31, 63, 313]. Therefore, energy storage is not necessary in these situations, but may help to protect the DC link of the converters from overvoltage.

As in the case of the fluctuation suppression service, the suitable storage systems for this application present high ramp-up rates, enabling a fast power modulation. Therefore, batteries, flow batteries, and short-timescale forms of energy storage such as supercapacitors, flywheels, and SMES systems are well suited for this application.

In Abbey and Joos [285], the DC link of the set of back-to-back converters of a wind turbine driving a DFIG is complemented by supercapacitors. Numerous simulation results show the improved ride-through capability of the system with energy storage support. Fuzzy logic control techniques are suggested to manage the interaction between the PCSs of the supercapacitors and the wind generator converter controllers, dumping the voltage variations of the DC link during these disturbances.

The use of these control theories is also proposed in Ali et al. [290]. This article deals with the implementation of SMES in a system with fixed-speed wind turbines, equipped with pitch control. The SMES system is connected to an AC cable through a six-pulse pulse width modulation (PWM) rectifier/inverter, using IGBTs and two-quadrant DC–DC choppers. Both converters are linked by a DC-link capacitor. The effectiveness of the pitch control and the SMES device in the voltage stability of the system in persistent fault situations, caused by the inability to re-close the circuit breakers of the system, is studied. The improvement of the voltage stability with SMES systems in LVRT situations is also discussed in Shi et al. [276].

Another PCS for SMES systems is presented in Kinjo et al. [274], comprised in this case of a combination of series and parallel inverters. Their DC links are connected to a two-quadrant DC–DC converter with a DC-link capacitor and a superconducting coil. While the series converter is responsible for regulating the voltage oscillations of the wind generator, the parallel converter simultaneously controls the active and reactive power in order to damp the oscillations of the tie-line power flow. The DC bus voltage is properly maintained by controlling the superconducting coil.

As well as SMES systems and supercapacitors, batteries and flow batteries are also proposed for LVRT applications. For instance, in Wang et al. [259], a VRB is connected to a DC link of a direct drive wind turbine driving a permanent magnet synchronous generator. The control of the DC–DC converter of the VRB enables an improved capability of the generator in LVRT situations.

7.2.3 Voltage Control Support

The control of the reactive power flow in an electrical network is crucial for maintaining proper voltage levels in the system. However, the control of active power flows can also enhance the performance of the voltage levels. Since they are connected through fully controllable power electronics, batteries, flow batteries, and short-timescale forms of energy storage such as supercapacitors, flywheels, and SMES systems can actively manage both active and reactive power, so they could provide excellent voltage control in the network.

Since the storage device must be able to manage both active and reactive power, the PCS of the storage device becomes essential. In this sense, FACTS/ESS systems are proposed to carry out this task properly; for example, Surive and Mercado [282] propose a distribution static synchronous compensator (DSTATCOM), coupled with a flywheel in order to mitigate voltage-stability problems in distribution networks due to the variable power output of renewable-based generators. Since the DC link of the STATCOM is strengthened by the energy storage support, it can exchange both active and reactive power.

It is important to note that active power control features depend on the storage technology. In this sense, a SMES system presents very good characteristics for fast injection or absorption of active power; for example, Hayashi *et al.* [270] show field test results for a SMES system, where a 16.6 ms response time in the step input of both active and reactive power can be seen.

7.2.4 Oscillation Damping

The electrical frequency stability in power systems depends on the so-called instantaneous power reserves of the system; that is, the inertia of synchronized generating systems. Any imbalance between power generation and consumption is instantaneously balanced due to the physical principle of the synchronous generator. The large inertia of the rotating generator set works as buffer storage, leading to the mentioned change in rotational speed and thus system frequency. The larger the synchronized inertia in the system, the slower is the change of frequency. The stepwise replacement of synchronized power plants by nonsynchronized renewable-based generating systems can affect the amount of the above-mentioned instantaneous power reserves, and thus the frequency stability of the system under disturbances.

For this reason, in future grid codes, wind parks will be required to help the generators of interconnected networks not to lose synchronism against perturbations. Thus, WPPs will be required to mitigate these power oscillations of the system by absorbing or injecting active power at frequencies of 0.5–1.0 Hz [252]. This way, they will somehow emulate the provision of instantaneous power reserves (the inertia) of synchronized generators.

Many storage technologies are suitable for this service. The time of injection/absorption of active power by the storage device is about 1 min; therefore high ramp-up rates and a fast response time are preferable. Thus, HESS, flow batteries, batteries, and short-timescale forms of energy storage such as supercapacitors, flywheels, and SMES systems are well suited for this application.

System stability aspects are usually dealt with by modal and frequency domain analysis. It is proposed that flywheels should be included in the network due to their better dynamic performance under disturbances [305, 306]. For instance, in Liu *et al.* [305], a case study of a general multimachine system is considered, in which a method for an optimal location for the installation of flywheel devices is examined in order to damp the low-frequency power oscillations of the system.

The capacity of an SMES system to quickly manage large quantities of active and reactive power simultaneously is investigated in Liu *et al.* [293], Ngamroo *et al.* [304], Padimiti and Chowdhury [294], and Wang *et al.* [299]. In these works, SMES is required to provide oscillation damping of power flows in an interconnected power system with renewables. A frequency domain analysis, based on linearized system models using eigenvalue techniques, as well as time domain analysis, based on a more detailed nonlinear system models in disturbance conditions, are proposed. These disturbances may be caused by the disruption of local loads, wind gusts, fast wind fluctuations, or short circuits. Control techniques are a key aspect here. Since system uncertainties such as various generating and loading conditions, parameter variations, and nonlinearities must be taken into account, the application of linear controllers is not always appropriate.

In this regard, it is interesting to note the methods described in Ngamroo *et al.* [304]. Here, the robust nonlinear control of a SMES system is proposed, which bases its operation on the addition of a power disturbance in a wind-based network with oscillating power flow in order to achieve a net constant power flow in the system. The consideration of the uncertainties of the system in SMES control provides an optimized performance.

Not only have theoretical studies been carried out, but also experimental tests [300]. Here, the benefits of the inclusion of storage devices to improve the stability of the system are discussed. It is concluded that power oscillation damping control is more robust against variations in power system conditions in the case in which the active and reactive power is managed by means of SMES systems and the actuation of batteries.

7.2.5 Primary Frequency Control

Power reserves can be defined as the additional active power (positive or negative) that can be delivered by a generating unit in response to a power imbalance in the network between generation and consumption. Three different reserve levels can be defined: primary, secondary, and tertiary reserves [48].

As presented in Chapter 3, the primary reserve is intended to be the additional capacity of the network that can be automatically and locally activated by the generators governor after a few seconds at most of an imbalance between demand and supply of electricity in the network [44]. Primary reserves must be delivered until the power deviation is completely offset by the other reserves in the network, which are called the secondary and tertiary reserves. The physical stabilizing effect of all connected rotating machines due to their inertia is not considered to be a primary reserve (on this, see Section 7.2.4).

Historically, electric power systems were characterized by holding vast amounts of power reserves. The majority of the generating sources, were synchronized with the network so that any mismatch between generation and demand could be compensated by synchronized and dispatchable plants based on gas or fossil fuels, thus ensuring the electrical frequency stability of the network.

However, the stepwise replacement of conventional generating units by nonsynchronized generating plants will have a significant impact on the frequency behavior of the system. First, it will lose the instantaneous power reserves mentioned in Section 7.2.4, because from the point of view of the system, wind or solar parks have no inertia; that is, the generator shaft is not connected directly to the system but is decoupled via fast controlled power electronics. Second, the grid will lose the active power reserves of conventional synchronized plants.

This is why future networks of renewable-based power plants will be required to provide primary reserves, just as conventional generating plants do. According to the definition of primary reserves, a wind park will be required to regulate its active power for up to 30 min in order to provide primary reserves for frequency support to the system.

There are many storage technologies that are suitable for regulating this power in a controlled manner to participate in services related to primary frequency support: flywheels, SMES systems, batteries, flow batteries, hydrogen-based energy storage systems (HESSs), compressed air energy storage (CAES), or pumped hydroelectric storage (PHS) installations.

The provision of primary reserves plays a key role, especially in isolated systems [308, 309]. In this sense, Mercier, Cherkaoui, and Oudalov [309] propose that battery energy storage systems (BESSs) should be included in an isolated wind–hydropower–gas system. The management, as well as the optimal size, of the batteries are the main concerns of study, in order to obtain the maximum economic benefit for the owner of the storage device while fulfilling the power reserve function. A numerical optimization problem is proposed in order to optimize the economic benefit, which is given by the difference between the revenues, due to the availability of frequency control reserves and income from sales of stored energy, and the costs, due to maintenance and investment in storage technologies.

The results from experience providing a spinning reserve using a 6 MW/6 MWh vanadium redox battery (VRB) in a 30.6 MW WPP are reported in Yoshimoto, Nanahara, and Koshimizu [258]. In conclusion, it is important to remark that wind turbine power output oscillations with a period around 30 min are reduced by a factor of three. The estimation of the battery charge state by means of cell voltage measurement favors the operation of the VRB. Flow batteries in spinning reserve applications have been reported extensively in the literature. In fact, their short response times and their capacity to be overloaded make these systems superbly well suited for this application, even to the extent of having advantages over other conventional facilities, such as fossil-fuel power stations [264].

7.3 An Example of Fluctuation Suppression: Flywheels for Wind Power Smoothing

As previously explained in this chapter, as ESS can be used to minimize the effects of the fast variability of renewable-based generating systems. Flywheel devices are effective for wind power smoothing. This example summarizes the findings in our previous work [212, 314] on wind power smoothing with flywheels. In particular, the research was focused on the optimal operation of the flywheel so as to smooth the net power injected to the grid by a wind turbine as much as possible, and on the design and testing of a practical controller for such a purpose, using laboratory-scale equipment.

7.3.1 The Problem of Wind Power Smoothing

Power fluctuations (in the time range of up to a minute) in wind turbines may cause fast voltage variations, especially in weak or isolated grids [265, 315]. In fact, and according to Sørensen, Hansen, and Carvalho-Rosas [316], Bianchi, De Batista, and Mantz [66], and Tascikaraoglu *et al.* [317], fast power fluctuations of wind turbines could markedly affect power quality levels. In particular, high flicker levels can be noted due to cyclic perturbations of the rotational torque, as well as other stochastic factors related to the randomness of the wind. A flicker is a voltage fluctuation that is clearly observed in lighting levels, while the so-called rotating sampling effect denotes to some extent the cycling torque perturbations due to the airflow deviation through the tower section each time a blade passes through it. The frequency of this perturbation is established by the mean rotational frequency of the turbine, P, for a given wind speed multiplied by the number of blades, N. Figure 7.1 shows a typical frequency–power spectrum corresponding to a three-bladed 1.5 MW wind turbine.

As can be seen, the concentration of energy is around the rotational frequency of the turbine, P, times its number of blades. The $1P$ component and its second harmonic $3P$ are clearly identified.

The present example focuses on the smoothing of the turbulent components of the power, which mainly corresponds to the rotating sampling effect. Only partial-load operation of the wind turbines is considered, as in the full-load operating region of the turbines, the variability of the power generated is alleviated by the action of the pitch actuator.

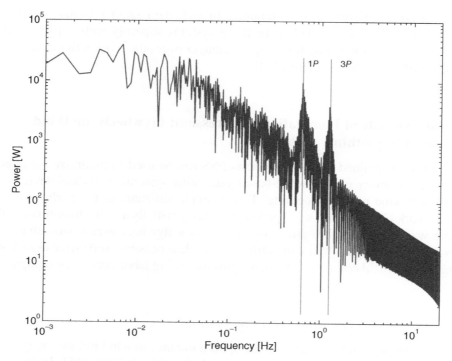

Figure 7.1 The frequency–power spectrum of a three-bladed 1.5 MW wind turbine in the partial-load operating region. *Source:* Díaz-González *et al.*, 2013. Reproduced with permission of Elsevier.

To summarize, Figure 7.2 depicts a conceptual diagram of the system under study. It comprises a variable-speed wind turbine and a flywheel-based storage device. The wind turbine provides highly variable power to the grid. The storage device exchanges power with an external network to smooth the power flow. For high wind power values, part of the energy is stored in the flywheel. Conversely, this energy is delivered to the grid during low wind power levels. Thus, the power injected into the grid is smoother – that is, less variable – than it would be if it was injected by a wind turbine without flywheel support. As can be observed, the flywheel should be governed by an energy management algorithm. The inclusion of this component is justified for the following reasons:

1. The limited energy capacity of the flywheel. Without an energy management strategy, the storage device would frequently become fully charged or discharged, thus limiting its operability.
2. The need to compensate the high-standing losses of the flywheel. Compensating the turbulent components of the wind power requires the flywheel to exchange

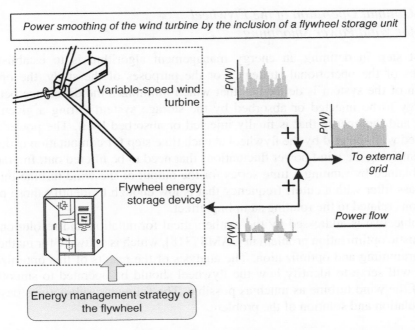

Figure 7.2 A conceptual diagram of a flywheel energy storage system (FESS) for wind power smoothing.

a power series with an average value close to zero. However, without proper compensation of the standing losses, the storage device will become continuously discharged.

3. The possibility of reducing the operational losses of the flywheel. The losses of the flywheel increase with the rotational speed. Thus, there is no reason to consider, for instance, high average rotational speeds if the flywheel is just intended to exchange low levels of power.

The design of the energy management algorithm comprises two main stages. In the first one, research is concentrated on the definition and determination of the optimal operation of the storage device for some given operating conditions for the wind turbine. The solution of this problem will be reached through the formulation and deterministic solution of an optimization problem in the General Algebraic Modeling System (GAMS). The second stage, based on the results obtained in the first one, focuses on the design of an energy management algorithm that can be implemented in a microprocessor to govern the charging and discharging of the storage device in real time. These two stages are deployed in the following sections.

7.3.2   Optimal Operation of the Flywheel
for Wind Power Smoothing

The first step in defining an energy management algorithm is to establish the objectives of the operational strategy. For the purposes of the study, the optimal operation of the system is defined as that which minimizes the difference between the energy to be injected or absorbed by the storage system during a given time interval, and the energy that is finally injected or absorbed by it. The power to be exchanged with the grid by the flywheel at each time step for computation is derived from identifying the wind power fluctuations that need to be filtered out. In practice, this is obtained by running a time series for the output of the wind turbine through a high-pass filter with a cutoff frequency that is low enough to identify those power fluctuations related to the rotating sampling effect.

This objective is addressed by the mathematical formulation and resolution of a deterministic optimization problem in GAMS [318], which is software for mathematical programming and optimization. The analysis of the results of this optimization problem will serve to identify how the flywheel should be operated to smooth the output of the wind turbine as much as possible. The following subsections describe the formulation and solution of the problem.

7.3.2.1   The Formulation and Solution of the Optimization Problem
for Operation of the Flywheel

The energy absorbed or injected into the grid by the flywheel over a certain time interval depends on two terms:

- The power due to the acceleration of the storage device, which is expressed as a function of its rotational speed, $\omega_{fw_t}^*$, and the electrical torque, T_{acc_t}:

$$P_{acc_t} = T_{acc_t} \omega_{fw_t}^*. \tag{7.1}$$

- The power losses of the storage device, which are expressed as a second-order function of its rotational speed:

$$P_{loss_t} = c_1 \omega_{fw_t}^{*2} + c_2 \omega_{fw_t}^*. \tag{7.2}$$

To smooth the power delivery to the grid, the flywheel has to compensate the reference power P_{ref_t} corresponding to the turbulent components of the wind power. Therefore, the difference ΔP between the power absorbed or injected by the flywheel and the reference power has to be minimized at all times. In terms of energy, this corresponds to minimizing the following function:

$$J = \sum_{t=1}^{N} \{u_t \underbrace{[P_{\text{ref}_t} - (P_{\text{acc}_t} + P_{\text{loss}_t})]}_{\Delta P} \Delta t - (1 - u_t) \underbrace{[P_{\text{ref}_t} - (P_{\text{acc}_t} + P_{\text{loss}_t})]}_{\Delta P} \Delta t\}, \quad (7.3)$$

where the binary variable u_t depends on the sign of P_{ref_t}.

Note that the difference ΔP, and J, would be zero only if the flywheel was able to compensate all the turbulent components of the wind power represented by P_{ref_t}. However, the energy and power that the storage unit is capable of injecting or absorbing are limited and depend on its state of charge (SoC). As previously noted, the power of the flywheel is bounded by the SoC, as it depends on the product of its rotational speed and its rated torque. Also, the energy that the storage device is capable of storing is bounded by the maximum rotational speed of the system; that is, by the maximum SoC.

As a consequence, ΔP would not be zero in practice and the optimal rotational speed of the flywheel $\omega_{\text{fw}_t}^*$ and the electrical torque T_{acc_t} must be obtained by solving the following optimization problem:

$$\min_{(T_{\text{acc}_t}, \omega_{\text{fw}_t}^*)} J, \quad (7.4)$$

subject to the following conditions:

- The storage device must remain within its speed operating limits,

$$\omega_{\text{fw}_{\min}} \leq \omega_{\text{fw}_t}^* \leq \omega_{\text{fw}_{\max}} \quad (7.5)$$

- The maximum electrical torque of the servomotor due to the acceleration of the system cannot be exceeded:

$$T_{\text{acc}_{\min}} \leq T_{\text{acc}_t} \leq T_{\text{acc}_{\max}}. \quad (7.6)$$

- The speed profile must satisfy the equation of motion of the system:

$$\frac{\omega_{\text{fw}_t}^* - \omega_{\text{fw}_{(t-1)}}^*}{\Delta t} k_2 = T_{\text{acc}_t}. \quad (7.7)$$

- The binary parameter u_t is set to 1 while positive values of P_{ref_t} are considered,

$$u_t P_{\text{ref}_t} \geq 0. \quad (7.8)$$

- The binary parameter u_t is set to zero while negative values of P_{ref_t} are considered:

$$(1 - u_t)P_{ref_t} \leq 0. \tag{7.9}$$

- The power injected by the storage device should not exceed the referenced power,

$$u_t P_{ref_t} \geq u_t(P_{acc_t} + P_{loss_t}). \tag{7.10}$$

- The power absorbed by the storage device should not exceed the referenced power:

$$(1 - u_t)P_{ref_t} \leq (1 - u_t)(P_{acc_t} + P_{loss_t}). \tag{7.11}$$

- The operating state of the storage device – that is, whether it is injecting or absorbing power – must agree with the sign of the referenced power:

$$P_{ref_t}(P_{acc_t} + P_{loss_t}) \geq 0. \tag{7.12}$$

The variables to be determined are as follows:

- T_{acc_t} is the electric current consumed or injected by the flywheel due to an acceleration or deceleration at time t.
- $\omega^*_{fw_t}$ is the reference angular speed of the flywheel at time t.

The input data are as follows:

- P_{ref_t} is the series of the reference power of the flywheel, obtained by computing the difference between the output power of the wind turbine and its filtered value at a given cutoff frequency. This reference power is also limited by the rated power of the flywheel. This series depends on the particular mean wind speed and its turbulence.
- u_t is a binary parameter that depends on the signal of P_{ref_t}.

The parameters of the optimization problem are as follows:

- ω_{min} and ω_{max}, in rad/s, are the operational speed limits of the flywheel.
- $T_{acc_{min}}$ and $T_{acc_{max}}$, in A, are the maximum positive and negative values for the electrical torque of the flywheel servomotor, due to the acceleration or deceleration of the system.
- k_2, in kgm^2, is a flywheel characteristic computed as $2J/p$.
- c_1, in Ws^2/rad^2, is a flywheel power losses characteristic.

- c_2, in Ws/rad, is a flywheel power losses characteristic.
- Δt, in seconds, is the unit time interval.

Given a time series of P_{ref_t}, the solution of the optimization problem determines the optimum instantaneous rotational speed and torque developed by the flywheel for the considered period of time. Thus, it provides the time series of the power exchanged by the flywheel so that the net energy injected and absorbed match the requirements as closely as possible, taking into account the limitations of the storage unit. This formulation results in a mixed-integer nonlinear problem, which is solved using GAMS.

7.3.2.2   Analysis of the Results

The optimization problem presented in the previous section is solved for sufficient representative cases (200 cases) to come up with consistent and generalized conclusions. The data needed for optimization are summarized in Table 7.3.

The adopted temporal trends in the wind power correspond to those for a DFIG-based wind turbine rated at 1.5 MW, exposed to variable wind speed profiles as described in Díaz-González *et al.* [212]. The reference power P_{ref} is computed by passing the wind power profile through a fourth-order Butterworth filter, with 0.4 Hz as the cutoff frequency. This filter allows us to separate out the power associated the the turbulence, which needs to be attenuated.

The solution of the optimization problem – that is, the temporal trends for the reference angular speed ω_{fw}^* and the instantaneous power exchanged with the grid by the flywheel, for a particular case – is depicted in Figure 7.3. In this figure, the time series of the reference power P_{ref}, and the actual power delivered by the flywheel, P_{fw}, for a particular wind profile are shown. The mean wind speed considered is 7.5 m/s

Table 7.3   The parameters of the system.

System	Parameter	Value
Storage device	P_{rated} (kW)	100
	ω_{max} (krpm)	31.0
	ω_{min} (krpm)	15.5
	$T_{\text{acc}_{\text{max}}}$ (Nm)	31.2
	$T_{\text{acc}_{\text{min}}}$ (Nm)	−31.2
	J (kgm^2)	0.72
Other parameters	k_2	0.36 kgm^2
	c_1	9×10^{-5} Ws^2/rad^2
	c_2	0.175 Ws/rad

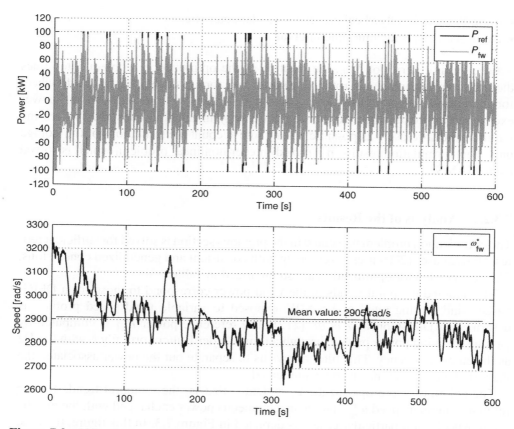

Figure 7.3 The reference power, the actual power delivered by the flywheel, and the optimal reference angular speed, ω_{fw}^*, corresponding to a wind profile of a 7.5 m/s mean wind speed and 0.05 pu of turbulence.

and the turbulence is 0.05 pu, which corresponds to 0.15 pu of wind power turbulence. A 600 s time interval is considered for each wind profile. It can be observed that 99% of the energy needed for the optimal solution can be provided by the flywheel. Recall that the reference power P_{ref} indicates the power that must ideally be provided by the flywheel to smooth the wind power and to take the power limitations of the storage device into account. The difference between P_{ref} and P_{fw} is a consequence of the SoC of the flywheel at a particular moment.

The analysis of the results imposing different wind profiles depict a dependency between the average angular speed of the flywheel and the average wind power for each case. Figure 7.4 presents the reference angular mean speed function of the mean wind power for a flywheel of 100 kW and a 1.5 MW wind turbine.

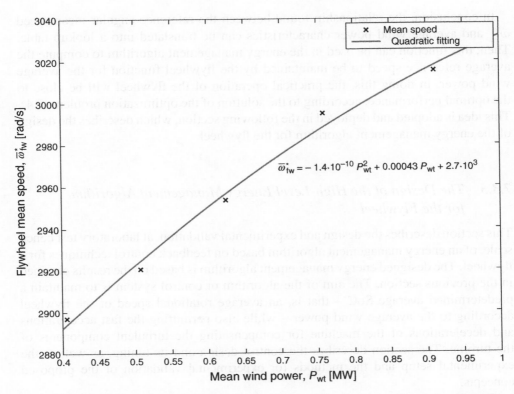

Figure 7.4 The relationship between the flywheel reference mean speed $\bar{\omega}_{fw}^*$ and the mean wind power obtained by analysing the optimal results. Each cross corresponds to the mean value of $\bar{\omega}_{fw}^*$ for all cases evaluated for each value of the mean wind power.

Only partial-load operation of the wind turbine has been considered. The results indicate that the higher the mean wind power, the higher is the mean rotational speed.

As shown in Figure 7.4, given wind turbulence, the wind power that the storage device has to compensate is higher considering high mean generation levels of the wind turbine than considering low generation levels (referring to the partial-load operation of the wind turbine), as most of the wind power is injected into the grid. The energy that the flywheel is able to exchange is proportional to the square of its rotational speed. Thus, it seems to be reasonable to consider different mean rotational speeds of the flywheel as being dependent on the mean expected energy levels to be injected or absorbed. This trend can be fitted by a second-order function. The reference mean speed is in the range of approximately 2900–3020 rad/s, which represents 89–93% of the SoC.

In conclusion, the relationship found between the reference angular mean speed $\bar{\omega}_{fw}^*$ and average wind power characteristics can be translated into a lookup table. Then, this function can be used in the energy management algorithm to compute the average reference speed to be maintained by the flywheel function for the average wind power. In doing this, the practical operation of the flywheel will be close to the optimal performance, according to the solution of the optimization problem (7.4). This idea is adopted and deployed in the following section, which describes the design of the energy management algorithm for the flywheel.

7.3.3 The Design of the High-Level Energy Management Algorithm for the Flywheel

This section describes the design and experimental validation, at laboratory test bench scale, of an energy management algorithm based on feedback control techniques for a flywheel. The designed energy management algorithm is based on the results obtained in the previous section. The aim of the algorithm or control system is to maintain a predetermined average SoC – that is, an average rotational speed of the flywheel according to the average wind power – while also permitting the fast accelerations and decelerations of the machine for compensating the turbulent components of the turbine. The section describes the control design and its tuning, as well as the experimental setup and the methods for experimental validation of the proposed concepts.

The energy management algorithm of the flywheel is presented in Figure 7.5. As can be noted, the algorithm consists of two main parts: the so-called inputs filtering and processing; and the feedback control. These two parts are described in the following subsections.

7.3.3.1 Inputs Filtering and Processing

The high-level energy management algorithm of the flywheel has two main objectives: to let the flywheel maintain an optimum average rotational speed, while enabling the fast accelerations and decelerations of the system. Mathematically, achieving these two objectives means tracking the reference average rotational speed $\bar{\omega}_{fw}^*$ and the high-frequency reference torque d^*. These signals are the setpoints entering the feedback control at different points, as can be seen in Figure 7.5. The block "inputs filtering and processing" in the energy management algorithm computes these setpoint signals from the input signals ω_{fw} and P_{wt}. The optimum average rotational speed $\bar{\omega}_{fw}^*$ depends on the mean wind power, as deduced in Section 7.3.2. In the present section, the dependence between these two variables is represented by the P–ω droop characteristic (see Figure 7.5).

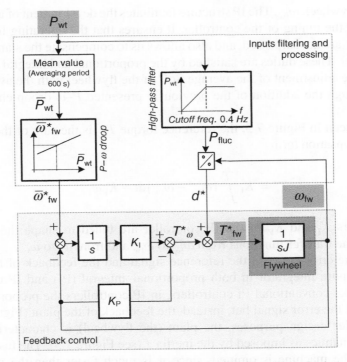

Figure 7.5 The energy management algorithm of the flywheel. Note the input signals of the algorithm (P_{wt} and ω_{fw}) and the output T^*_{fw}.

While the optimum average rotational speed $\bar{\omega}^*_{fw}$ is a slowly varying signal, the reference torque d^* contents high-frequency components, and is computed as follows:

$$d^* = \frac{P_{fluc}}{\omega_{fw}}, \tag{7.13}$$

where P_{fluc} corresponds to the fast wind power fluctuations that must be attenuated. This signal is obtained by passing the wind power measurements through a high-pass filter. The cutoff frequency of this filter is chosen so that the power disturbances caused by the rotating sampling effect pass through the filter without significant distortion.

7.3.3.2 Feedback Control

The aim of the feedback control is to ensure the tracking of the slowly varying reference $\bar{\omega}^*_{fw}$ and the fast varying reference d^*. To this end, an integral–proportional (IP) controller acts on the reference torque T^*_ω to regulate the instantaneous rotational speed of the flywheel, ω_{fw}. The IP structure facilitates the development of a systematic procedure for the tuning of the controller. It ensures that the operating limits of the storage device are not surpassed, and also allows us to compensate the standing losses of the flywheel. These duties are satisfied by the proportional and integral parts of the controller. The adjustment of the average SoC of the flywheel with the average wind power is through the addition of the previously presented P–ω droop characteristic (see Figure 7.5).

As can be seen in Figure 7.5, the reference torque T_{fw}^* is the sum of the torque d^* and the compensation term

$$T_{\text{w}}^* = K_{\text{I}} \int \left(\bar{\omega}_{\text{fw}}^* - \omega_{\text{fw}} \right) \mathrm{d}t - K_{\text{P}} \omega_{\text{fw}}. \tag{7.14}$$

The parameters K_{I} and K_{P} of the IP controller are tuned to shape the frequency responses of the transfer functions from $\bar{\omega}_{\text{fw}}^*$ to ω_{fw} and from d^* to ω_{fw}.

Note that the error between the reference signal and the feedback of the plant is affected by a pure integrator in both proportional–integral (PI) and IP controllers. However, unlike conventional PI controllers, in IP controllers the proportional gain does not affect the error signal but, instead, the feedback of the plant (Figure 7.5).

For controller design purposes, the plant (the flywheel) is characterized by its mechanical dynamics as imposed by the inertia J (see Figure 7.5). The fast electrical dynamics of the machine is omitted, since it is much faster than the mechanical dynamics. Accordingly, the flywheel speed is given by

$$\omega_{\text{fw}}(s) = \frac{1}{Js + K_{\text{P}}} \left(d^*(s) + \frac{K_{\text{I}}}{s} \left(\bar{\omega}_{\text{fw}}^*(s) - \omega_{\text{fw}}(s) \right) \right). \tag{7.15}$$

Then, rearranging terms,

$$\omega_{\text{fw}}(s) = T_{d^*\omega}(s)d^*(s) + T(s)\bar{\omega}_{\text{fw}}^*, \tag{7.16}$$

where

$$T_{d^*\omega}(s) = \frac{s}{Js^2 + K_{\text{P}}s + K_{\text{I}}}, \tag{7.17}$$

$$T(s) = \frac{K_{\text{I}}}{Js^2 + K_{\text{P}}s + K_{\text{I}}}. \tag{7.18}$$

The reference d^* is not a completely exogenous signal, since it depends on the flywheel speed according to (7.13). To consider this fact in the parameter tuning and to guarantee stability at all possible values of the power and flywheel speed, the reference is expressed as

$$d^* = \delta \times \omega_{fw}, \tag{7.19}$$

where $\delta = P_{fluc}/\omega_{fw}^2$ is a time-varying parameter that takes values in the interval $[-\delta_{max}, \delta_{max}]$, with

$$\delta_{max} = \frac{P_{max}^*}{\omega_{min}^2}. \tag{7.20}$$

With the previous definitions and using the small gain theorem [319], it is possible to state conditions to ensure the stability of the closed loop system for all acceptable values of the power and the rotational speed. More precisely, the closed loop system is stable for all values of δ in $[-\delta_{max}, \delta_{max}]$ if the infinite norm of the transfer function from d^* to the output of the plant ω_{fw} $(T_{d^*\omega})$ does not exceed the upper limit δ_{max}; that is,

$$\|T_{d^*\omega}\|_\infty = \max_\omega |T_{d^*\omega}(j\omega)| < \delta_{max}. \tag{7.21}$$

It is assumed that the transfer function $T_{d^*\omega}$ has two real and different poles; that is,

$$T_{d^*\omega} = \frac{s/K_I}{(s/p_1 + 1)(s/p_2 + 1)}$$

$$= \frac{s/K_I}{\dfrac{1}{p_1 p_2}s^2 + \left(\dfrac{1}{p_1} + \dfrac{1}{p_2}\right)s + 1}. \tag{7.22}$$

If p_1 is the dominant pole $(p_1 \ll p_2)$, then

$$\|T_{d^*\omega}\|_\infty < \frac{p_1}{K_I} < \frac{1}{\delta_{max}}. \tag{7.23}$$

Therefore, the integrator gain K_I should satisfy

$$K_I > p_1 \delta_{max} \tag{7.24}$$

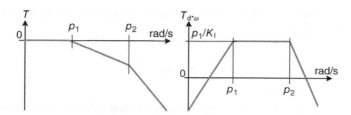

Figure 7.6 The asymptotic diagram of the frequency responses of the transfer functions $T_{d^*\omega}$ and T.

to ensure closed-loop stability for all possible values of the power and speed. Furthermore, the proportional gain can be obtained from comparing the denominators of (7.17) and (7.22):

$$K_P = K_I \left(\frac{1}{p_1} + \frac{1}{p_2} \right), \tag{7.25}$$

where $p_2 = K_I/Jp_1$.

In Figure 7.6, an asymptotic graph of the frequency responses of $T_{d^*\omega}$ and T can be observed. The pole p_1 defines the bandwidth of the speed tracking as well as the infinity norm of $T_{d^*\omega}$. Therefore, once the bandwidth of T is known, the parameters of the controller can be computed from (7.24) and (7.25).

7.3.4   Experimental Validation

This section describes the flywheel test bench and the wind turbine emulator, as well as the rest of the laboratory equipment used for configuring the system for the purposes of the study. Emphasis is placed on presenting those actions needed to emulate the variability of the power of the wind turbine using laboratory-scale equipment, among other considerations. Then, the experimental results are analysed with the aim of validating the proposed energy management algorithm of the flywheel.

7.3.4.1   A Description of the Experimental Setup

Figure 7.7 presents a scheme of the experimental setup. As can be noted, the system is composed of a flywheel test bench, a wind turbine emulator, a coupling transformer that connects the system to the grid, measurement devices, and communication and control devices. Each of these main components of the system is detailed in what follows.

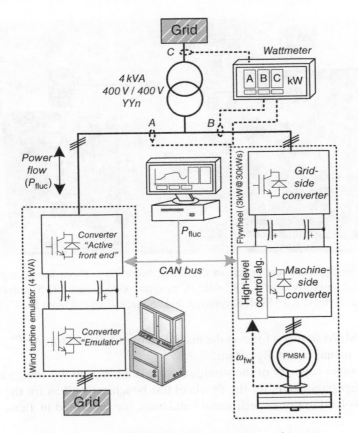

Figure 7.7   The scheme of the experimental setup.

The Flywheel Test Bench

The flywheel test bench is composed of a rotating disk, mechanically coupled to the shaft of a permanent magnet synchronous machine (PMSM). The electrical machine is controlled by a set of back-to-back power converters. These power converters are driven by digital signal processor (DSP)–based control boards (see Figure 7.8). The system is connected to the grid through a coupling transformer.

As previously noted, the flywheel is added at the point of connection of the wind turbine emulator. The design of the current control loops for the power converters of the flywheel are explained in depth in Chapter 6 and in Díaz-González *et al.* [250]. To summarize, the grid-side converter of the system is in charge of regulating the voltage of the DC link of the back-to-back power converters and also the reactive currents exchanged with the network. The control of the machine-side converter is the field-oriented vector control system of the PMSM [213, 246]. This control algorithm

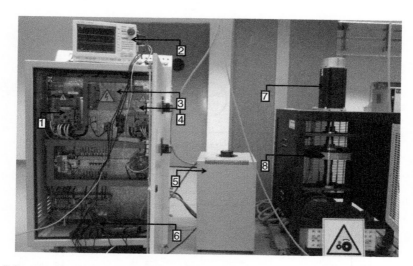

Figure 7.8 The flywheel test bench in the IREC laboratory. From left to right: 1, grid-side converter; 2, oscilloscope; 3, DC link; 4, machine-side converter; 5, autotransformer; 6, measurement devices; 7, PMSM; 8, rotating disk. *Source* Adapted from IREC, 2015 [108].

governs the PMSM so that it follows the instantaneous electrical torque T_{fw}^* referenced by the energy management algorithm.

The rated power capacity of the storage device is 3 kW, and the energy capacity is 30 kWs. Further parameters for the flywheel test bench, as well as for the rest of the equipment involved in this experimental validation, are presented in Table 7.4.

The Wind Turbine Emulator

The wind turbine is emulated by a cabinet consisting of two identical three-phase voltage sources in back-to-back configuration [320] (see Figure 7.9).

A bidirectional power flow through the converters is possible, since they can be operated as either active rectifiers or active inverters. According to Figure 7.9, while representing the behaviour of a generator, power flows from the AC side of the "emulator" converter to the AC side of the converter "active front end". The reverse process depicts the behavior of a load. The low-level control algorithms of the power converters of the cabinet receive the series of active and reactive power setpoints in order to represent the power profile of the wind turbine. The rated apparent power of the cabinet is bounded to 4 kVA. Further details of the cabinet are presented in Ruiz-Álvarez *et al.* [320].

Measurement Devices

For the purposes of the chapter, it is necessary to analyze the variability of the power injected by the emulator of the wind turbine, and its attenuation due to the inclusion of the flywheel. Accordingly, a wattmeter simultaneously registers the power series at

Table 7.4 The parameters of the experimental setup.

System	Parameter	Value
Wind turbine emulator	Two three-phase IGBT bridges back-to-back	
	S_{rated} (kVA)	4.0
	$U_{ACrated}$ (V)	400.0
	$U_{DCrated}$ (V)	750.0
	I_{max} (A)	16.0
	C (F)	0.0050
Flywheel	Ratings	3.0 kW @ 30 kWs
	Efficiency	73%
	ω_{max} (rpm)	3000.0
	ω_{min} (rpm)	1000.0
	T_{rated} (Nm)	12.2
	J (kgm^2)	0.868
Flywheel (PMSM)	ψ_m (Wb)	0.2465
	L_d & L_q (H)	2.88×10^{-3}
	R_s (Ω)	0.44
	Pole pairs	2
Flywheel (power electronics)	(Same as emulator)	

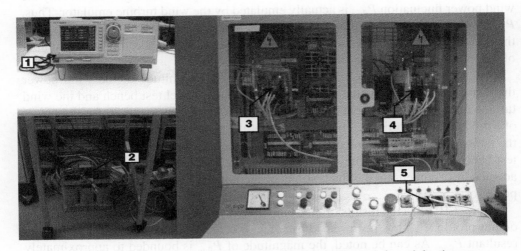

Figure 7.9 The experimental setup in the IREC laboratory. From left to right: 1, wattmeter; 2, coupling transformer; 3, power converter of the wind turbine emulator "active front end"; 4, "emulator" power converter; 5, CAN bus port. *Source* Adapted from IREC, 2015 [108].

the points A, B, and C depicted in Figure 7.7 for their post-processing and analysis. These points corresponds to the terminals of the wind turbine emulator, the terminals of the flywheel test bench, and the point of connection of the system to the network, respectively.

Communications

As presented in Figure 7.7, the wind turbine emulator represents the variability of the power of the wind turbine following the power setpoints sent through a controller area network (CAN) bus by a computer. The CAN bus is also used to let the energy management algorithm of the flywheel know the above-mentioned wind power. The precise clocking of the digital signal processor (DSP) of the control board of the machine-side power converter of the flywheel is used as a time basis to coordinate the exchange of signals between the computer, the wind turbine emulator, and the flywheel. In particular, the signals are sent through the CAN bus each $20 \, \mu s$.

7.3.4.2 Assumptions for the Emulation of the Fluctuating Components of Wind Power

According to Figure 7.5, the average optimum rotational speed of the flywheel $\bar{\omega}^*_{fw}$ and also the wind power fluctuations P_{fluc} have to be determined by filtering the power generated by a wind turbine, P_{wt}. However, the presented experimental validation avoids the emulation of the power P_{wt} and consequently its filtering. Instead, only the wind power fluctuation P_{fluc} is actually emulated by the wind turbine emulator. Thus, P_{fluc} is a power profile with an average value close to zero, which is obtained from the simulation of the system and the application of the high-pass filter to the output of the model of the wind turbine P_{wt}.

This is carried out with the aim of adjusting the magnitude of the wind power fluctuations P_{fluc} to the actual power ratings of the flywheel test bench and the wind turbine emulator. The present chapter considers a scaling factor of 20 for the power of a 1.5 MW wind turbine. The magnitude of the resultant profile, P_{fluc}, is similar to the rating of the wind turbine emulator (4 kVA), and also to the rating of the flywheel test bench (3 kW). Therefore, the size of the flywheel test bench can be considered adequate to compensate the fluctuations of the power of a wind turbine with a rated power of 1.5/20 MW.

Figure 7.10 plots the results of the simulation of a 1.5 MW wind turbine. Its power output profile, P_{wt}, has been scaled by a factor of 20. The subplot below shows the resultant P_{fluc}. As can be noted, the magnitude of P_{fluc} is bounded to approximately 3 kW, which corresponds to the rated power of the flywheel test bench. The cutoff frequency of the high-pass filter of P_{wt} is set to 0.4 Hz, so that the rotating sampling effect can be represented.

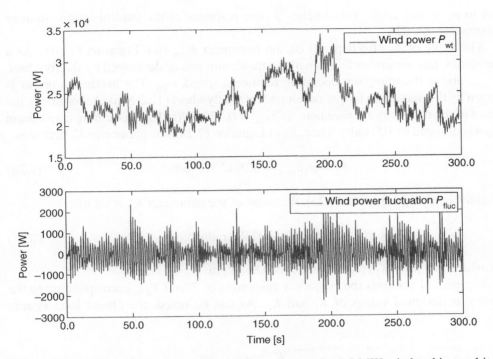

Figure 7.10 The scaled magnitude of the power output of a 1.5 MW wind turbine and its fluctuating components with a cutoff frequency of 0.4 Hz.

As a consequence of the direct emulation of P_{fluc}, the P–ω droop characteristic (Figure 7.5) is not applied. Therefore, the optimum average rotational speed of the flywheel, $\bar{\omega}^*_{\text{fw}}$, will be step-profiled in the following sections to evaluate the performance of the designed feedback control.

7.3.4.3   The Determination of the Control Parameters for the Experimental System

Section 7.3.3.2 depicts the procedure for the tuning of the parameters of the IP structure that builds up the feedback control. This section presents the particular values of the controller determined for the study case.

As can be noted in Equations (7.24) and (7.25) the parameters of the IP controller, K_P and K_I, depend on the location of the pole p_1, which in turn bounds the value of the pole p_2. The pole p_1 determines the time response of the control loop for the reference $\bar{\omega}^*_{\text{fw}}$. This time response can be very slow, since the dynamics of $\bar{\omega}^*_{\text{fw}}$ depends on the average value of P_{wt}, with an averaging period of 600 s. Thus, the pole p_1 is set to $p_1 = 0.01$ rad/s. This implies a time response of the control loop system of approximately 628 s.

The value of K_I also depends on the parameter δ_{max} (see Equation (7.20)). As a reminder, this parameter is given at the maximum power developed by the flywheel, P_{max}, and at the minimum operating rotational speed, ω_{fw}. The maximum power is given by the maximum torque developed by the flywheel (12.2 Nm). Accordingly, the limit of the time-varying parameter is $\delta_{max} = 0.122$ W/(rad/s)^2, assuming a minimum flywheel speed of 100 rad/s. Then, from Equation (7.24), the parameter K_2 becomes

$$K_I = p_1 \delta_{max} = 0.0012 \text{ W/(rad/s)}. \tag{7.26}$$

Finally, applying Equation (7.25), the value of the parameter K_P is set to

$$K_P = 0.1307 \text{ W/(rad/s)}^2, \tag{7.27}$$

provided that $p_2 = K_P/(J \times K_1)$ and J is 0.868 kgm^2.

Figure 7.11 presents the frequency responses of T and $T_{d^*\omega}$ corresponding to the previous designed values of K_I and K_P. As can be noted, the closed loop system

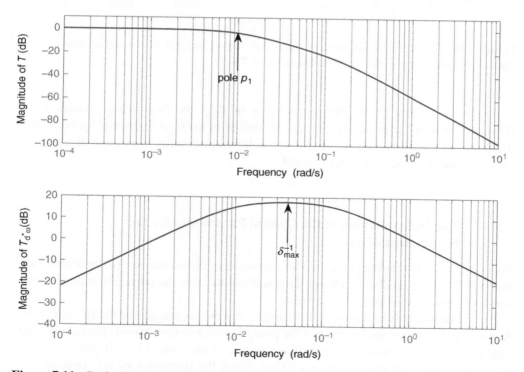

Figure 7.11 Bode diagrams of the closed loop transfer function T (from $\bar{\omega}_{fw}^*$ to ω_{fw}) and the transfer function $T_{d^*\omega}$ (from d^* to ω_{fw})

will be able to track reference speeds until the cutoff frequency of 0.01 rad/s, which corresponds to the location of pole p_1. The frequency response of $T_{d^*\omega}$ shows that the infinity norm is below the limit $(\delta_{max})^{-1} = 8.19$ Nm/(rad/s) (18.3 dB). The graph also shows that the system is able to track reference torques ranging from 0.01 to 0.1 rad/s with the maximum gain allowed by the stability guarantee and the flywheel torque limits.

7.3.4.4 Analysis of the Experimental Results

In this section, the experimental results obtained to evaluate the performance of the proposed energy management algorithm of the flywheel are presented. As noted in previous sections, the wind turbine emulator reproduces the fluctuating power components depicted in Figure 7.10, which the flywheel test bench is in charge of compensating. Accordingly, Figure 7.12 depicts the actual power developed by the wind turbine emulator, the flywheel, and the net power exchanged with the network.

As can be noted, the flywheel compensates the fluctuating components of the wind power, leaving the net power profile almost constant. However, the average value of the power of the flywheel, and thus of the net power exchanged with the network, is

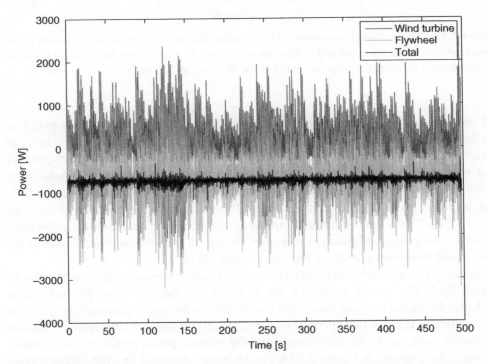

Figure 7.12 The power of the wind turbine emulator, the flywheel, and the net power exchanged with the network. The average rotational speed of the flywheel is $\bar{\omega}_{fw} = 220$ rad/s.

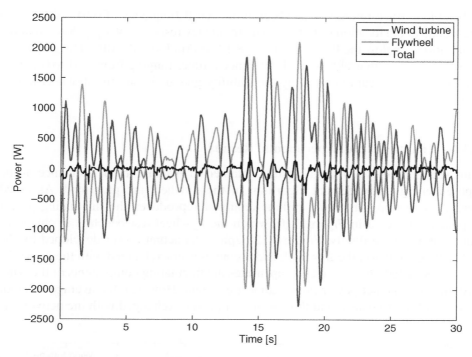

Figure 7.13   The instantaneous power of the wind turbine emulator (blue line), as well as the power profiles of the flywheel and at the network terminals, subtracting the standing losses of the flywheel. The average rotational speed of the flywheel is 220 rad/s.

not zero due to the need to compensate the power losses of the flywheel to maintain the indicated average rotational speed $\bar{\omega}_{fw}^*$. In this case, the optimal average rotational speed is close to 220 rad/s [212]. As this setup is meant to be a proof-of-concept system, the losses in the flywheel test bench are much higher than in a commercial flywheel storage device. In particular, the energy efficiency of the test bench is 73%, and the power losses depends on the rotational speed, reaching up to 800 W at the rated speed [250]. These figures are far from those corresponding to a high-tech flywheel, for which the energy efficiency is around 90% and the level of the power losses at the rated speed represent just 2% of the rated power [321].

   Figure 7.13 examines more closely the previously presented power profiles in Figure 7.12. The power profile of the flywheel and the profile of the net power exchanged with the network have been corrected by subtracting the standing losses of the flywheel at the constant speed of $\bar{\omega}_{fw} = 220$ rad/s. In this way, the average value of the power exchanged with the network and the average power profile of the flywheel are zero. As a result, in Figure 7.13 it can be better observed that the instantaneous power of the flywheel compensates to a great extent the fluctuating components of the

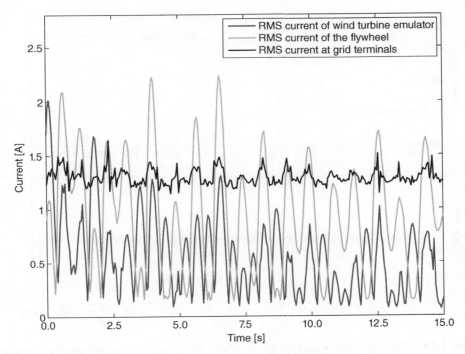

Figure 7.14 The RMS electric currents of the wind turbine emulator, the flywheel, and the network. The average rotational speed of the flywheel is $\bar{\omega}^*_{fw} = 220$ rad/s.

power of the wind turbine. This instantaneous regulation of the power of the flywheel is governed by the reference torque d^* (see Figure 7.5).

The RMS currents of the flywheel, the wind turbine emulator, and the network are depicted in Figure 7.14. As shown, the magnitudes of the currents of the flywheel are much higher than those of the wind turbine emulator, due to the need to compensate the losses of the system, as previously discussed. Without the proper loss compensation, the storage device will become completely discharged.

Apart from compensating the fast fluctuations of the power of the wind turbine, the second objective of the feedback controller of the flywheel is to maintain a determined average rotational speed $\bar{\omega}^*_{fw}$. The performance of this control loop is depicted in Figure 7.15. As can be observed, the flywheel is rotating steadily at an average speed of 220 rad/s and the time response to a step-profiled average reference speed $\bar{\omega}^*_{fw}$ from 220 to 270 rad/s is around 600 s, as imposed by the design of the controller presented in Section 7.3.4.3. Moreover, in the subplot below it can be observed that the voltage of the DC link of the flywheel test bench remains stable at an average of 750 V while regulating the rotational speed of the flywheel and compensating the fast wind power fluctuations.

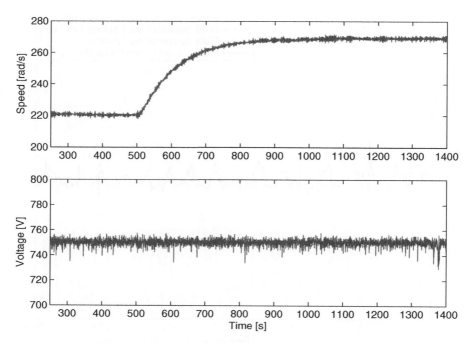

Figure 7.15 The rotational speed of the flywheel in response to a step-profiled average reference speed $\bar{\omega}^*_{fw}$ from 220 to 270 rad/s. The DC-link voltage of the flywheel test bench is presented in the bottom subplot.

The power spectrum of the wind turbine emulator, as well as the total power exchanged with the network when supporting the flywheel, are depicted in Figure 7.16. The average rotational speeds of the flywheel are 120 and 220 rad/s. As can be noted, the wind turbine emulator clearly represents the rotational sampling effect of the turbine at approximately 0.6 and 1.2 Hz. The rotating sampling effect is mostly compensated by the flywheel support. The constant component of the power spectrum of the flywheel due to its losses has been subtracted, so that the performance of the system can be better observed. This figure depicts the support of the flywheel, considering differing average values of the SoC. It is worth noting that the support that the flywheel can provide is better while rotating at an average speed of 220 rad/s (close to the optimum) than while rotating at 120 rad/s. This is because the power capability of the flywheel is bounded by the product of the speed and the rated torque of the electrical machine and thus, as discussed in previous sections, there is an average optimal speed for the flywheel, $\bar{\omega}^*_{fw}$, which is dependent on the magnitude of the fluctuating components of the wind power to be compensated.

The performance of the system, considering differing average flywheel operating rotational speeds, is quantified from the attenuation of the fluctuating components of the net power exchanged with the network. This magnitude is computed from the data

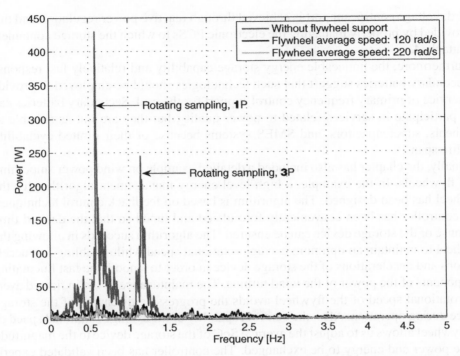

Figure 7.16 The spectrum of the net power exchanged with the network without flywheel support, as well as with a flywheel at different average values of the SoC. The average power losses of the flywheel have been subtracted.

shown in Figure 7.16. The ratio between the energy of the fluctuating components of the net power exchanged with the network considering the flywheel support (red and green lines in Figure 7.16), and the fluctuating components of the wind turbine emulator (blue line in Figure 7.16), gives the attenuation. An attenuation of 85% is found while the flywheel is rotating at an average speed of 120 rad/s. The attenuation reaches 92% with 220 rad/s as the average rotational speed of the flywheel.

7.4 Conclusions

This chapter has shown that the fast response, high ramp power rates, and high cyclability of short-term storage systems such as, principally, flywheels, supercapacitors, and SMES systems can be exploited to provide various services in the power system in regard to power quality improvement. For instance, such short-term storage installations can be used to smooth the power output of renewables, thus facilitating their integration into the grid.

Other potential applications are those related to voltage control support, also including eventualities in the power system such as short circuits and voltage sags. In this regard, voltage control support is achieved through reactive power regulation and this is provided by properly managing the electronic PCSs to which the storage containers are attached.

Furthermore, the noticeable energy storage capability and relatively fast response of secondary batteries suggest that such technologies should be employed to provide the service of primary frequency control support for the grid. Secondary batteries can also participate in services related to power quality, but they are not as suitable as flywheels, supercapacitors, and SMES systems because of their limited cyclability and life span.

Finally, the chapter has also included a detailed example on wind power smoothing with flywheels. In the example, a high-level energy management algorithm for the flywheel has been designed. The algorithm is based on feedback control techniques. The controller has been conveniently formulated and tuned so that the desired time response of the storage device can be ensured. The algorithm succeeds in allowing the flywheel to maintain an optimum average rotational speed while enabling fast accelerations and decelerations of the storage device in order to smooth the fast fluctuating components of the power of the wind turbine. The maintenance of the indicated average rotational speed of the flywheel avoids the progressive discharge of the storage device during operation. Also, the control capability of the average rotational speed of the flywheel allows us to adjust the average SoC of the storage device to the magnitude of the power and energy to be exchanged. The controller has been validated experimentally in laboratory-scale equipments. The results show that most of the fluctuating components of the power of the wind turbine due to the rotating sampling effect can be compensated through flywheel support.

8

Mid- and Long-Term Applications of Energy Storage Installations in the Power System

8.1 Introduction

To complete the catalog of storage applications started in Chapter 7, this chapter describes potential mid- and long-term applications that energy storage systems (ESSs) could provide in the electric power system. In general terms, the inclusion of energy storage installations in the power system is motivated here by the need to ensure the required balance between generation and demand considering mid- and long-term (in the range of hours) timescales, regardless of the variability of renewable energy sources. Since the electric power systems of the future will be characterized by holding increasing and important penetration rates of renewables, most of the applications for storage installations discussed hereinafter will be closely related to renewable generation. As in Chapter 7, each technical issue has been identified and defined according to Barton and Infield [160], Bayod-Rújula [251], Beaudin et al. [166], Dell and Rand [135], EPRI [252], Georgilakis [253], and Świerzyński et al. [254]. In addition, the definition of these aspects is complemented by a brief discussion on the role of the ESS in each case.

First, and with the aim of providing an initial general picture, the mid- and long-term services that EESs can provide in the power system are listed and classified by storage technology in Tables 8.1 and 8.2. The literature cited in these tables mostly relates EESs to wind power (for further details, see our previous work [42]).

8.2 A Description of Mid- and Long-Term Applications

8.2.1 Load Following

In this service, storage technologies are required to provide energy in a time frame ranging from minutes up to 10 h [160]. Due to the stochastic nature of renewables,

Energy Storage in Power Systems, First Edition. Francisco Díaz-González, Andreas Sumper and Oriol Gomis-Bellmunt.
© 2016 John Wiley & Sons, Ltd. Published 2016 by John Wiley & Sons, Ltd.

Table 8.1 An overview of publications regarding the uses of ESSs in the field of wind power (part I).

	Storage duration at full power	PHS	HESS	CAES	VRB	ZBB	PSB
Load following	Min–10 h	[322–326]	[262,308,327–332]	[333–335]	[127,260,336,337]	[260,337–339]	[260]
Peak shaving	1–10 h	[150,322,340–343]	[328,329,344,345]	[260,346]	[337,347,348]	[311,337,339]	[348]
Transmission curtailment	5–12 h	[91,340,349,350]	[328,351,352]	[346,350,353]	[354]	$\sqrt{}^a$	$\sqrt{}^a$
Time shifting	5–12 h	[349]	[90,329,345]	[355]	[252]	[311]	
Unit commitment	Hours–days	[322]	$\sqrt{}^a$	[263]			$\sqrt{}^a$
Seasonal storage	≥ 4 months	[92]	[330,352]				

^{a}Although the storage technology is suitable for this application, dedicated studies are not listed here.

Table 8.2 An overview of publications regarding the uses of ESS in the field of wind power (part II).

	Storage duration at full power	NaS	Lead–acid	Ni–Cd	Li-ion
Load following	Min – 10 h	[101, 337, 356–358]	[296, 310, 337, 358, 359]	[267, 337, 358]	[358]
Peak shaving	1–10 h	[154, 337, 358, 360]	[296, 337, 358, 361]	[337, 358]	[358]
Transmission curtailment	5–12 h	$\sqrt{}^a$			
Time shifting	5–12 h	[337]			
Unit commitment	Hours–days				
Seasonal storage	≥4 months				

^{a}Although the storage technology is suitable for this application, dedicated studies are not listed here.

the plant output would not match the power demand. This leads to various technical and economic problems regarding the operation of the electrical system. Technical issues, such as voltage and frequency variations due to imbalances between electricity generation and demand, limit the penetration of renewable technologies into the electrical network. With regard to economic issues, it should be remarked that some regulatory frameworks specify economic penalties for operators that do not meet their generation bids on account of wind and solar forecasting errors. In this sense, the ESS can be used to store and inject electric power for hours. Batteries and flow batteries, as well as hydrogen-based energy storage systems (HESSs), compressed air energy storage (CAES), or pumped hydroelectric storage (PHS) installations are well suited for this application. Figure 8.1 graphically depicts the concept of load following with storage support.

A prominent example of the feasibility of combining wind with battery solutions is probably the case of a wind power installation in Futumata (Japan), where a 34 MW NaS battery bank is used to level the production of a 51 MW wind power plant (WPP) [357]. Proper management of the battery energy is essential, not only with regard to technical issues (e.g., shortages/surpluses of the battery) but also from an economic point of view. In this sense, in Hida *et al.* [356], a control algorithm that optimizes the economic benefit of the system, minimizing the storage in times of peak demand when the market price of the energy is high, is developed.

In this case, control and dimensioning aspects of flow batteries are discussed in Barote and Marinescu [336], Barote *et al.* [127], and Brekken *et al.* [338]. To

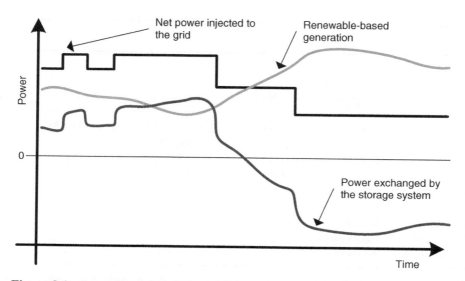

Figure 8.1 A graphical description of the concept of a load-following application.

summarize these works, it can be said that many techno-economic benefits for the electrical system derive from a proper solution of these aspects. Proper control of the batteries improves the predictability of renewable-based plants and, therefore, the associated costs for their integration into the grid inasmuch as reserve requirements are concerned can decrease, since great precision can be achieved in matching their output to their forecast power. According to Brekken *et al.* [338], 34 MW and 40 MWh of storage capacity are required to improve the forecast power output of a 100 MW wind plant (34% of the rated power of the plant) with a tolerance of 4%/pu, for 90% of the time.

Techno-economic analyses are addressed in Cavallo [333], Lund and Salgi [125], and Swider [335], regarding the use of CAES in load-following applications. As an example, Swider [335] presents a stochastic electricity market model in order to study the effects of high penetration of wind power in electrical systems, as well as the economic viability of including CAES solutions. With the minimization of system costs as the criterion, there is an optimization problem that takes into account aspects such as transmission capabilities of the system, energy prices, the technical characteristics of the generating plants, electricity demand profiles, investment costs, and power reserve requirements. Important conclusions (taking the German electricity market into account), include the economic advantages of CAES over conventional peak thermal plants in a scenario with a high penetration of wind power.

Finally, it is important to remark that hydrogen-based storage technologies are considered as one of the most promising technologies in load-following applications. Actually, several demonstration projects have been developed as a proof of concept concerning stand-alone systems with wind, photovoltaic (PV) generation, and hydrogen storage [327, 329, 330]. These projects focus on developing power management algorithms, using the excess of energy to create hydrogen in an electrolyzer and then using it in a fuel cell in order to inject power to the system when required. The evaluation of the operation of the system shows the technical feasibility of such isolated schemes with hydrogen support.

8.2.2 Peak Shaving

This service falls within a time frame ranging from 1 to 10 h. The operating strategy for the storage devices is to store cheap energy during off-peak hours (overnight), and to inject it into the network during periods of high electricity demand, hence softening the typical mountain and valley shape of the load curve.

Well-suited ESSs for peak-shaving applications are batteries, flow batteries, CAES, HESS technologies, and PHS. Regarding the batteries, numerous techno-economic studies display the feasibility of storing energy during off-peak hours and selling it in periods of peak demand.

In Dufo-López, Bernal-Agustín, and Domínguez-Navarro [337] and Roberts [360], the use of NaS batteries for this application is discussed. While technical benefits for the electrical system in a real case, as well as details referring to the design of the power conversion system (PCS) of the battery, are presented in Roberts [360], an interesting techno-economic analysis of a battery energy storage system (BESS) is discussed in Dufo-López, Bernal-Agustín, and Domínguez-Navarro [337]. In conclusion, in order to define an available economic operation of a BESS in the Spanish energy market, the sale price of the battery energy is fixed at (€0.22–0.31)/kWh [actually, the energy price is around (€0.04–0.05)/kWh]. Therefore, it is concluded that BESS operators should receive subsidies, due to the emissions that would result from the use of conventional fossil-fuel plants for peak-shaving applications, in order to make the use of a BESS economically profitable.

The selling price of BESS energy is substantially lower than that of a regenerative fuel cell (RFC) system. According to Bernal-Agustín and Dufo-López [344], to make a RFC economically viable for operation with a WPP would imply fixing its energy selling price at €1.71/kWh in the Spanish case, due to the low energy efficiency of the storage technology and the high cost of its components. Therefore, compared with the selling price of energy injected by batteries, the selling price of energy injected by hydrogen-based technology is between five and eight times higher. This is one of the main challenges regarding the inclusion of hydrogen-based storage systems in the network.

Without a doubt, PHS is considered to be one of the most well-suited storage systems in order to achieve high penetration levels of wind power in isolated systems. Indeed, wind–hydropower systems have been studied in, amongst other publications, Brown, Peças Lopes, and Matos [322], Kapsali and Kaldellis [342], and Papaefthimiou *et al.* [343]. A techno-economic study of the viability of wind–hydropower systems in providing power during periods of peak load demand is performed in Kapsali and Kaldellis [342]. The results show excellent technical and economic performance. It can be concluded that the integration of WPPs in the isolated study case can be increased by 9%, allowing for a penetration level of 20%. In addition, a significant reduction of CO_2 emissions through the use of PHS installations rather than using fossil-fueled peak power plants is highlighted in Benitez, Benitez, and Van Kooten [150]. However, regarding the dynamic security issues of operation of the system, it can be concluded that it may be appropriate to add some further technologies in order to provide a spinning reserve for the system [322].

8.2.3 Transmission Curtailment

In this application, storage technologies are required to provide energy in a time frame of 5–12 h. For a number of reasons, such as the need to ensure the stability of the electrical system or technical limitations in power transmission lines, renewable-based plants have to be disconnected. In this sense, an ESS can store energy for hours and inject it in a controlled manner, according to the capacity of the transmission lines and the resolution of stability issues, thus avoiding the disconnection of renewables. Well-suited ESS for this application are flow batteries, CAES, hydrogen-based systems, and PHS installations.

Studies regarding wind–hydropower systems and CAES installations for transmission curtailment applications are considered by Anagnostopoulos and Papantonis [340], Denholm and Sioshansi [353], Dursun and Alboyaci [91], and Zafirakis and Kaldellis [346]. In general, wind-based isolated systems or systems connected to weak grids are considered to display the most interesting scenarios. Research findings concur with the idea of including ESS in highly renewable penetration systems, with the aim of reducing wind curtailment, ensuring backup power, minimizing transmission losses, ensuring security of supply, saving updating costs, and avoiding the building of new transmission lines.

Finally, since hydrogen can be created by means of rejected wind power, HESS technologies are considered promising for inclusion in power system applications. Once the hydrogen is stored, it can be used in various ways: either to generate electricity in fuel cells and inject it into the network during periods of peak power demand or for other uses, such as in the field of mobility. As mentioned in the previous section, the main challenges for the inclusion of HESSs are related to the uncertainty of their economic viability (owing to their high system costs and low energy efficiency) and the dependence on high market prices for hydrogen [328, 350–352].

8.2.4 Time Shifting

In time-shifting services, storage technologies are required to provide energy in a time frame of 5–12 h. In this case, an ESS is required to absorb all the energy from renewable-based plants during off-peak hours, supplemented by cheap power bought from the network if necessary, and to sell it during periods of peak demand, thus avoiding the activation or updating of other conventional peak-power generation plants.

Flow batteries, CAES, PHS installations, and HESS technologies are well suited for this application.

In Nyamdash, Denny, and Malley [355], the effects on the operation of electrical networks of considering bulk energy storage capacity and WPPs are discussed. In this sense, many operating strategies for wind ESSs are considered. One of the most interesting study cases is based on charging the storage device continuously for a 12 h period (low-demand period) and injecting its power in a controlled manner during the following 12 h (high-demand period). In conclusion, it is highlighted that time-shifting services by means of the inclusion of an ESS in the network are not economically viable without some kind of subsidy, due to the high investment costs of the technologies (in this case, CAES systems are the most favorable technology) and the relatively low energy efficiency (depending on the technology). With regard to environmental aspects, an ESS should be able to inject power during the entire period of high peak demand; otherwise, the operation of base-load plants would be increased, with a consequent increase in CO_2 emissions.

8.2.5   Unit Commitment

In unit commitment services, storage technologies are required to provide energy in a time frame ranging from hours to days. Due to the uncertainties regarding mesoscale variations of the wind and of solar irradiation, it is hard to manage the commitment of wind turbines and solar panels in order to meet the estimated demand at all times. Also, the introduction of renewables into electrical systems motivates the need to maintain a certain level of energy reserves in order to compensate forecast errors. Therefore, the introduction of a high-capability ESS into the network may be useful to combat the effects of uncertainties in wind forecasting and to reduce the energy reserves of the system during its normal operation. Large-scale ESSs are suitable for this application: CAES and PHS installations, as well as HESS technologies.

This topic is addressed as a numerical optimization problem, in which the objective function is to minimize the operational costs of the electrical network so as to maximize the return on investment through the inclusion of an ESS [263, 322]. For instance, in Daneshi *et al.* [263], the unit commitment problem is formulated in a power system with wind generation and CAES. The benefits of including CAES solutions – in order to reduce the operational costs of the electrical network by means of allowing the use of wind energy in charging this storage technology when the energy is not needed by the system, and thus avoiding the disconnection of the wind turbines – are discussed.

8.2.6   Seasonal Storage

In this application, ESSs capable of storing and injecting energy during periods in a time frame of months are well suited. The storage of energy for long periods of time can be useful in systems with large seasonal variations in the level of generation or consumption. Clearly, only those storage technologies with a very large energy capacity and no self-discharge are eligible, such as large PHS installations or hydrogen-based solutions.

In cases where it can be technically interesting to include seasonal storage, and taking into account the investment costs regarding the installation of wind turbines and storage systems based on hydrogen, it may seem favorable to oversize WPPs in order to reduce the size of the storage reserves [352]. However, this would increase the range of the nonutilized wind power capacity and hence decrease the efficiency of the system. On the other hand, the energy costs of the system would be reduced.

A demonstration project regarding seasonal storage by means of HESS technologies in a stand-alone system is described in Little, Thomson, and Infield [330]. It must be noted that although the storage of energy for long periods of time is technically feasible due to the absence of leaks in the hydrogen storage tank, the use of the RFC must be limited, in order just to store the excess productions of wind power, in favor of minimizing the losses of the system, since the energy efficiency of RFCs is very low.

8.3   Example: The Sizing of Batteries for Load Following in an Isolated Power System with PV Generation

This section presents an example on battery sizing for isolated power systems with renewable generation. The scope of the study is to derive a rough first battery dimensioning as a function of the main characteristics of the isolated power system that it is part of, and also from basic operational assumptions. The final dimensioning of the storage system requires additional detailed calculations that are not included here. The sizing procedure presented here is adapted that presented in "IEEE standard 1013-2007: IEEE Recommended Practice for Sizing Lead–Acid Batteries for Stand-Alone Photovoltaic (PV) Systems" [362], and in "IEEE standard 1562-2007: IEEE Guide for Array and Battery Sizing in Stand-Alone Photovoltaic (PV) Systems" [363]. As a difference with the proposed methodology in the above-mentioned standards, the present scope includes PV generation profiles as a decision factor for battery dimensioning. Also, the procedure tackles the determination of the number of power inverters needed for grid connection of the storage system.

For the purposes of the study, the power system layout in Figure 8.2 is adopted. As can be noted, the battery bank is based on m strings of n battery cells, so as to achieve the required ratings. The required DC voltage ratings at the inverter terminals will determine the n battery cells connected in series. Also, the connection of the battery bank with the rest of the system will comprise p inverters – which means that it may require the parallel connection of inverters – and this number will depend on

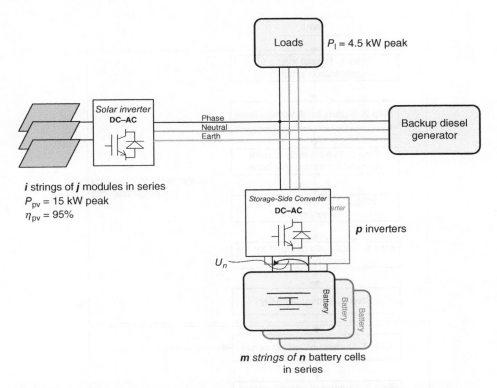

Figure 8.2 The layout of an isolated power system with PV generation and storage.

the determined power requirements for the battery bank and the ratings of commercial inverters. The power generated from the PV arrays and the power exchanged by the battery bank will be translated to the load – a residential building – through a single-phase AC system. To complete the system, a backup diesel generator is included to ensure the power supply to the load in the event of a prolonged scarcity of PV generation.

The aim of the battery-based storage system is to continuously compensate the mismatches between PV generation and load demand. Therefore, in terms of the mid- and long-term applications for storage installations identified in Section 8.2, the battery bank will provide the service of load following. The system, however, will be dimensioned so as to have enough energy storage capacity to supply the load during a few days without any PV generation. After this period, which is to be defined in the sizing procedure, the power supply to the load is supposed to be ensured by the backup diesel generator.

The sizing procedure comprises several phases and these are graphically summarized in Figure 8.3. As indicated, the battery dimensioning process starts with an evaluation of typical load and generation profiles. From such an evaluation, the procedure continues with the determination of the battery energy storage capacity, also

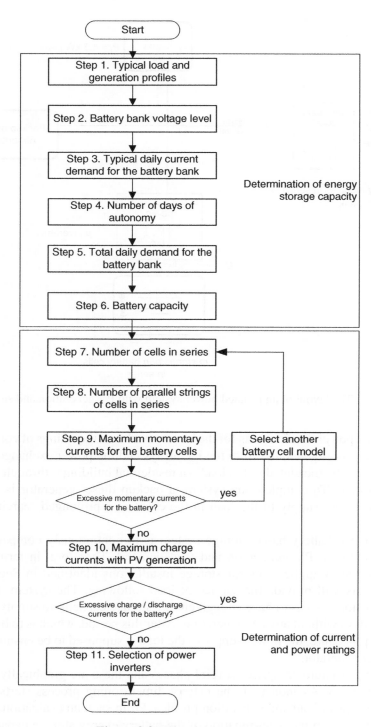

Figure 8.3   The sizing procedure.

considering technical operational constraints such as the maximum depth of discharge (DoD). Finally, the number of cells in series and parallel building up the battery bank, as well as the required number of power inverters for grid connection, complete the sizing procedure. The successive phases of the procedure is set out in the following sections.

8.3.1 Step 1: Typical Load and PV Generation Profiles

The first step is to evaluate the load and PV generation profiles for the residential building and the PV modules of the system, since these will serve to weight the power demand profile for the storage system. To this end, 12 typical PV generation and consumption daily profiles, with a 1 h time resolution, are adopted. Each of the 12 datasets represents a typical day for each of the months of the year. Such information represents the average behavior of the system in normal operating conditions over a whole year, and thus it proves very convenient in establishing a benchmark or reference scenario for battery dimensioning.

Figure 8.4 presents 12 typical daily load profiles, each one being representative of one of the months of the year. As can be noted, the power demand for the house reaches a peak of almost 4.5 kW on a typical morning in January.

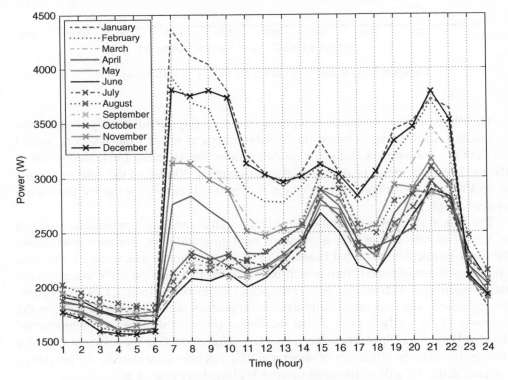

Figure 8.4 Typical daily load profiles for each of the months of the year.

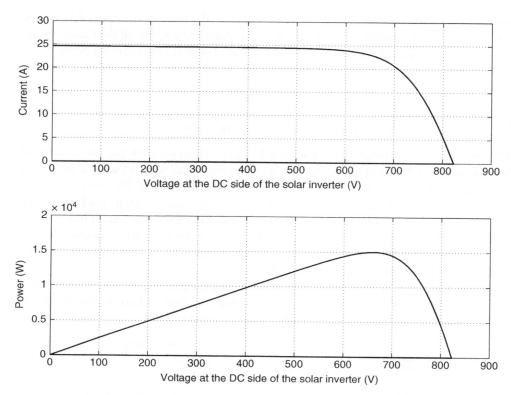

Figure 8.5 *I–V* and *P–V* curves for the PV generating system.

The PV generating system is comprised of three strings of solar arrays or modules in parallel. Each string is based on 25 solar arrays in series from Kyocera, model KC200GT. The parameters for the module can be found online [364, 365]. Each of the modules provides a peak power of approximately 200 W, so the system is rated at 15 kW peak (see Figure 8.5, in which *P–V* and *I–V* curves for the PV generating system are plotted).

The typical PV power generation profile for each of the days representative of each of the 12 months of the year is obtained through simulation. For each simulation, the daily profile for the global irradiation and temperature to which the PV modules are exposed are obtained from a web-based interactive map on a PV geographical information system from the JRC (European Commission) [366]. This tool offers, among others, site-dependent daily irradiation profiles with a 1 h resolution. For the purposes of the study, typical global irradiation (horizontal plane) and temperature data for Barcelona (Spain) have been adopted. After deducting the losses in the solar inverter (an average efficiency of about $\eta = 95\%$ has been considered), the net power injected at the AC side of the solar inverter is plotted in Figure 8.6.

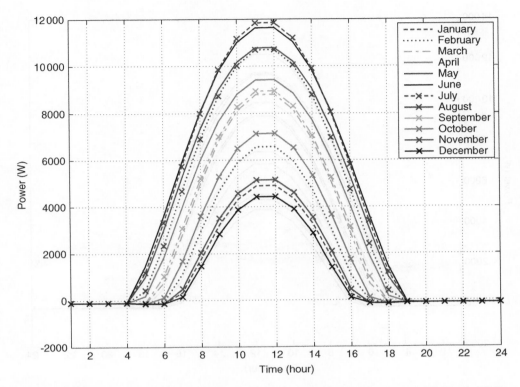

Figure 8.6 Typical daily PV generation profiles. The number accompanying each of the months in the legend sorts the months per magnitude of peak PV power generation.

As can be noted, and from the adopted irradiation and temperature conditions, the expected PV generation levels would be around 12 kW at maximum.

Finally, and to sum up, Figure 8.7 synthetically compares typical PV generation and load profiles. As can be observed, the expected PV generation levels could greatly exceed the peak power demand. However, the expected yearly energy yield by the PV generating system proves to be insufficient during five months of the year (from October to February), as presented in Table 8.3. During spring and summer, the PV energy surplus could serve, for instance, to feed other loads, but this is not addressed in the present example.

8.3.2 Step 2: The Voltage Level of the Battery Bank

By subtracting the typical daily load profiles from the typical PV generation profiles, we can deduce the typical or averaged daily power demand profiles for the storage system for each of the months of the year. These power demand profiles are key to sizing the batteries. However, the ratings of the batteries, both in terms of energy and

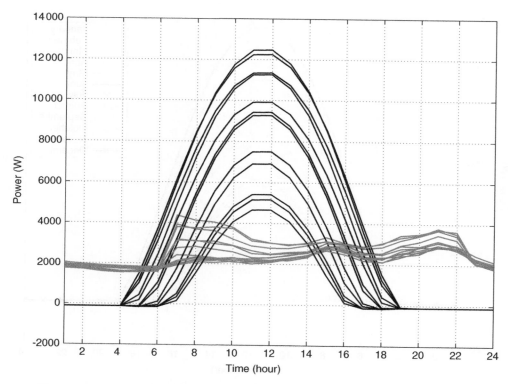

Figure 8.7 A comparison between typical daily PV generation and load profiles.

Table 8.3 A comparison between the yearly energy yield of the PV generating system and the expected yearly energy consumption.

Month	Generation (kWh)	Consumption (kWh)	Energy balance (kWh)
January	882	2092	−1210
February	1317	1985	−677
March	2031	1840	191
April	2399	1716	683
May	2953	1641	1311
June	3267	1587	1679
July	3248	1653	1595
August	2778	1735	1043
September	2112	1629	482
October	1518	1637	−118
November	960	1796	−835
December	764	1044	−1279

power capacities, should not be determined directly from the power profiles, but from the current profiles to be exchanged by the battery bank. To obtain these profiles, one could directly divide the power demand profiles by the nominal voltage of the battery bank. In this way, the determination of the nominal voltage U_n proves to be unavoidable in this early phase of the sizing methodology.

The power inverter manufacturer SMA provides specific guidelines to determine the battery bank voltage function of the ratings of the attached power inverter. For the "Sunny Island" product family [198], the manufacturer recommends to set U_n at around 48 V DC while configuring storage systems rated between approximately 4 and 30 kW. The power ratings of the storage facility to be included in the isolated power system under study will surely be within the above-mentioned range, so we shall set $U_n = 48$ V DC.

8.3.3   Step 3: The Typical Daily Current Demand for the Battery Bank

Once the battery bank voltage has been determined, the typical daily current profiles for the battery bank are calculated, and these are presented in Figures 8.8 and 8.9.

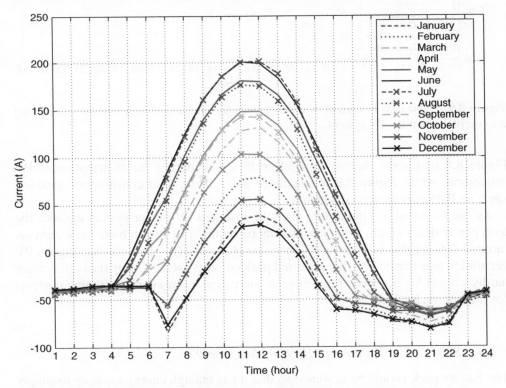

Figure 8.8 Typical daily current demand profiles for the battery bank considering the contribution of the PV system.

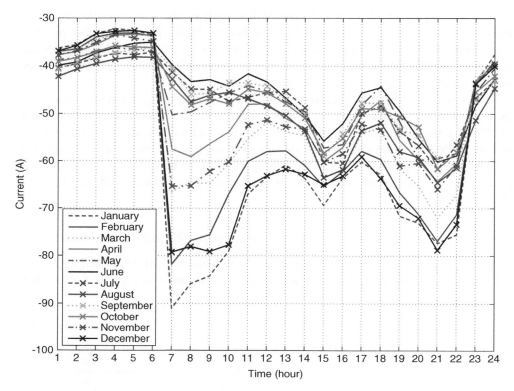

Figure 8.9 Typical daily current demand profiles for the battery bank without considering the contribution of the PV system.

Figure 8.8 represents current demand profiles considering the contribution of PV generation, while Figure 8.9 is calculated solely from load consumption profiles, since no PV generation is considered.

As can be noted in Figure 8.8, since the installed PV power system exceeds the load power demand, the battery bank may be required to absorb huge peak currents (around 200 A) due to excessive generation. Conversely, while considering no PV generation (see Figure 8.9), the expected peak currents exchanged by the battery bank hardly exceed 90 A. Here, these currents are called the maximum running currents for the battery, $I_{\text{max\_r}}$.

8.3.4 Step 4: The Number of Days of Autonomy

The battery pack should be designed so that it has enough energy capacity to supply the loads of the system (i.e., the residential building) for a few days without taking PV generation into consideration. The number of days of autonomy needs to be defined by the designer. After this period, the power supply would be ensured by the diesel generator. For the present example, the number of days of autonomy, d, is limited to 3 d.

8.3.5 Step 5: The Total Daily Demand for the Battery Bank

The total daily demand for the battery bank is computed here, as it will be used later for the calculation of the required energy storage capacity. The calculation is performed by simply integrating over time the daily current profiles in Figure 8.9. The results are presented in Figure 8.10.

As can be noted, since no PV generation is considered, the required capacity for the battery increases monotonically (the negative sign in the capacity is due to the fact that, by agreement, currents entering the battery are negative). Another important assertion is that the required capacity depends on the month. The worst-case scenario, for which the required capacity a maximum, is considered to be the typical load demand for January. In this case, the required capacity, Q, reaches 1433.3 Ah.

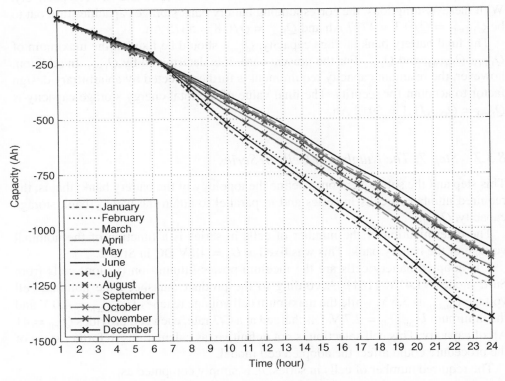

Figure 8.10 Typical daily current demand profiles for the battery bank without considering the contribution of the PV system.

8.3.6   Step 6: The Capacity of the Battery

Once the required daily capacity for the battery has been determined, it is quite straightforward to calculate the final dimensions of the bank. As previously defined, for backup operation, the battery should be able to feed the load for $d = 3$ d. In this way, the energy storage capacity of the battery bank should be at least $Q \times d = 4300$ Ah.

However, the batteries cannot be fully discharged during normal operation, since they will reach the exponential cell voltage zone while attaining the full discharge condition. Such an exponential trend in the cell voltage is not convenient for operation of the power inverter to which the battery bank is connected. As a consequence, the maximum DoD for the battery should be limited, further oversizing the capacity of the battery. Assuming a maximum DoD for lithium-ion battery cells of about 80%, the required battery bank energy storage capacity increases up to $Q_{DoD} = Q/0.8 = 5375$ Ah.

According to the standard [362], other criteria apart from the maximum admissible DoD could be considered for adjusting the required battery capacity. These are the maximum admissible daily DoD, DDoD, and the capacity of the battery at the end of life, EoL. In the present case, we consider these to be around 30 and 80%, respectively. With these assumptions, the corresponding battery bank storage capacities turn out to be $Q_{DDoD} = Q/0.3 = 4777$ Ah and $Q_{EoL} = Q/0.8 = 5375$ Ah.

The final battery bank storage capacity, Q_{total}, should be fixed at the maximum of Q_{DoD}, Q_{DDoD}, and Q_{EoL}. For the present case, it is determined by Q_{DoD}. In addition, however, the resultant capacity requirement is further affected by an arbitrary design factor to account for security. The final value of the total energy storage capacity is $Q_{total} = Q_{DoD}/1.1 = 5912$ Ah.

8.3.7   Step 7: The Number of Cells in Series

This step and the subsequent ones define the topology of the battery bank; that is, the required number of cells in series and in parallel to attain the battery bank storage capacity, Q_{total}.

The determination of the number of cells in series is a function of the nominal voltage of the battery bank. This was fixed at $U_n = 48$ V DC in Step 2.

The adopted battery cells are the medium-power lithium-ion (Li-ion) cells from Saft, model VL41M [161]. According to the product datasheet, the nominal cell voltage U_{n_cell} is 3.6 V, while the maximum cell voltage reaches $U_{max\_cell} = 4.0$ V and the minimum $U_{min\_cell} = 2.7$ V. Discharged at a C/3 rate, each cell provides $Q_{cell} = 41$ Ah of capacity. This cell selection is not definitive, as calculations in further steps of the procedure could affect the adopted cell model.

The required number of cells in series, n, is simply computed as

$$n = U_n/U_{n\_cell} = 48/3.6 = 13.3 \quad \rightarrow \quad 13 \text{ cells.} \tag{8.1}$$

Doing this, the minimum battery cell voltage will drop down to $n \times U_{\text{min\_cell}} = 35.1$ V and the maximum voltage will reach $n \times U_{\text{max\_cell}} = 52$ V. These values are set to be in accordance with the battery pack battery management system (BMS) voltage operating limits.

8.3.8   Step 8: The Number of Parallel Strings of Cells in Series

The attainment of the previously determined battery bank energy storage capacity requires us to connect various strings m in parallel, each composed of n cells in series. This number is determined by

$$m = \frac{Q_{\text{total}}}{Q_{\text{cell}} \times n} = \frac{5912}{533 \times 13} = 11.09 \quad \rightarrow \quad 11 \text{ strings.} \tag{8.2}$$

8.3.9   Step 9: Check the Admissible Momentary Current for the Battery Cells

The maximum currents for the battery bank, obtained in Step 3 of the design procedure, do not take possible peaks in load demand into consideration. Such peaks could be representative, for instance, of motor-starting processes. These are to be considered in the battery sizing procedure. Labeled as $I_{\text{max\_p}}$, in the present example such peak currents are weighted as four times the maximum running current for the battery due to the load demand. Therefore, $I_{\text{max\_p}} = 363.9$ A.

Since the cells are arranged in $m = 11$ strings in parallel disposition, each of the strings will provide a fraction of the above-presented maximum momentary current. Finally, the maximum current per battery cell will turn out to be

$$I_{\text{max\_cell}} = I_{\text{max\_p}}/m = 33.1 \text{ A.} \tag{8.3}$$

According to the battery cell datasheet, each cell could withstand up to 300 A for 30 s, so the maximum expected momentary current does not represent a constraint for the system as it stands.

8.3.10   Step 10: The Maximum Charge and Discharge Currents for the Battery Bank Considering PV Generation

At this point in the methodology, the battery bank is fully determined, both in terms of capacity and the number of cells in series and in parallel. The following last steps of the methodology serve to check the suitability of the selected battery cells in terms of the maximum charge and discharge currents that they should withstand in normal operating conditions. Note the difference between the maximum charge and discharge currents, or "running currents" hereinafter, and the admissible momentary currents determined in Step 9.

So far, the battery bank dimensioning procedure has not included the PV generation profiles as affecting factors. However, and as presented in Figure 8.8, the maximum running currents in the battery while storing excess PV generation could even double the maximum running current requirements for the battery associated with the load demand (see Figure 8.9).

Moreover, if the battery were to be sized so that it could store all excess PV generation at the current ratings presented in Figure 8.8, it would have to be unreasonably oversized both in terms of energy and power capacity. This is because, as presented in Table 8.3, there are months in which the PV generation exceeds the total energy consumption, and so this excess generation will be directly translated to the battery bank, thus enormously increasing the required number of battery cells. In this regard, Figure 8.11 presents the cumulative capacity for the battery bank resulting from storing all excess PV generation. Positive values of capacity at the end of the day denote a net increment in battery charge, while negative values indicate a net decrement in

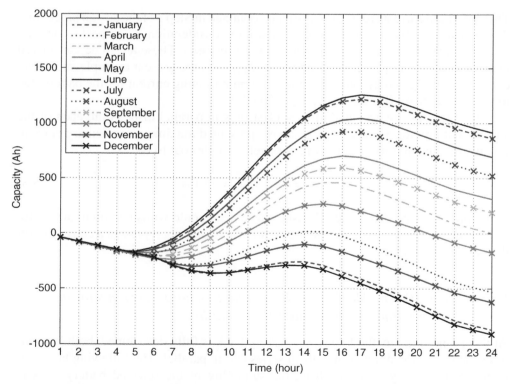

Figure 8.11 The charge accumulated by the battery bank while storing as much PV generation as possible.

battery charge. Those trends presenting a negative cumulative capacity at the end of the day correspond to typical days from October to February, and these are precisely the months in which the net energy balance denotes a deficit. Conversely, during spring and summer, the energy balance proves to be positive, and this is denoted by the positive cumulative capacity at the ends of typical days in Figure 8.11.

Therefore, it is important to limit the amount of excess PV energy that is stored and also the current ratings at which this energy is stored, so as not to incur unreasonably oversized systems. In the present example, we consider that the net energy balance between the energy generated and the energy absorbed by the battery bank for each typical day should equal zero (or should be negative, in the event of experiencing less PV generation than the power demand from the load). In this way, we will prevent the battery bank from storing, mainly during summer and winter periods, more PV energy than is actually needed for the battery to compensate mismatches between PV generation and demand. Doing this, the daily cumulative charge for the battery during a typical day for each of the 12 months of the year is presented in Figure 8.12.

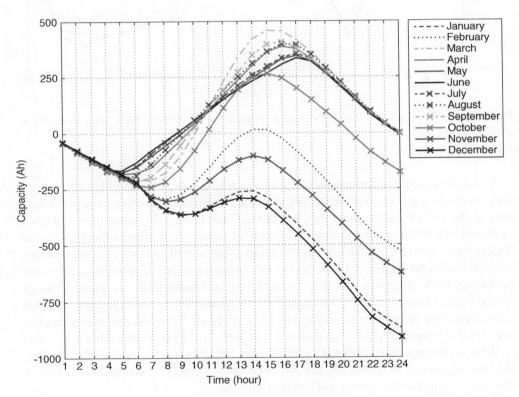

Figure 8.12 The charge accumulated by the battery bank while storing as much PV generation as needed.

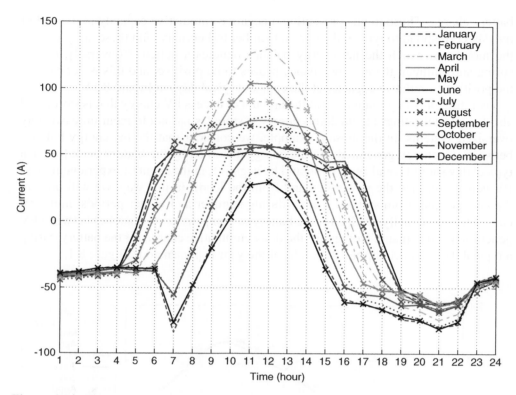

Figure 8.13 Typical daily current demand profiles for the battery bank with limited PV energy storage.

As can be noted, for spring and summer months, in which the monthly PV energy yield exceeds the consumption energy demand, there is no positive net charge balance at the end of each typical day, but the balance is zero. This has been obtained by iteratively limiting the maximum admissible running currents exchanged by the battery bank, until the admissible current ratings that actually ensure a zero net charge balance are found. Figure 8.13 shows the determined maximum running currents for the battery bank. As can be seen, for the particular case under study, the current ratings for the battery bank turn out to be around 130 A, which corresponds to the maximum current entering the battery at a typical midday in March. Since the net energy balance between PV generation and load demand is negative for March (see Table 8.3) – that is, there is not enough PV generation to feed the monthly load demand – the admissible running current in the battery has not been limited and this exceeds the maximum limited currents for the spring and summer months.

The previous calculations show that depending on the PV generation levels, the maximum running current flowing through the battery bank ($I_{\text{max\_rpv}}$, around

130 A) could exceed the maximum running current $I_{\text{max\_r}}$ if just the load power profile (around 90 A; see Step 3) is taken into consideration, thus neglecting the impact of PV generation.

Therefore, it is necessary to check the suitability of the selected cell model to withstand the running currents. The maximum running current $I_{\text{max\_rpv}}$ will be provided by $m = 11$ parallel strings of cells, so the current through each of the strings, and thus through each of the series-connected cells, will be around $I_{\text{max\_rpv}}/m = 11.8$ A. According to the battery cell datasheet, the maximum admissible running current is around 150 A, so the selected battery cells suit the current requirements.

8.3.11   Step 11: The Selection of Power Inverters

The last step of the design procedure is to select the number and ratings of the power inverters attached to the battery bank. As for the rest of the procedure, the determination of the final topology of the PCS for the battery bank will undoubtedly require further detailed calculations that are not included here.

In Step 2 of the methodology, the voltage level on the DC side of the converter was fixed at $U_n = 48$ V. Therefore, the question now is to determine the number p of parallel inverters so that the net power installed is high enough to charge and discharge the battery bank at the calculated ratings.

As deduced in the previous step, the maximum running current for the battery bank was $I_{\text{max\_rpv}} = 130$ A. Consulting the datasheet for SMA's "Sunny Island" family of power inverters [198] again, model 6.0H admits up to 100 A, which corresponds to 4.6 kW at nominal DC voltage (i.e., $U_n = 48$ V). So we need to include at least $p = 2$ power inverters in parallel to fulfill the required current ratings.

8.4   Conclusions

This chapter has presented different mid- and long-term applications for storage installations in electric power systems. The literature review has shown that the technical characteristics (in terms of energy capacity, power capacity, lifetime, and scalability, among others) of secondary batteries and flow batteries, PHS, CAES, and also HESS technologies, define these systems as well suited to ensure the required balance between generation and demand, regardless of the variability of the output of renewables.

Apart from providing such a balancing service, storage installations can maximize the usage of renewables by avoiding the disconnection of such power plants as a result of technical limitations in power transmission lines. In addition, storage can time-shift the output of renewables to periods of high demand, thus avoiding the connection of gas or fossil-fuel power plants for this purpose.

To summarize, mid- and long-term storage installations can be defined as potential drivers for the decarbonization that the power system is currently experiencing, and will also experience in the mid-term. However, the implementation of these technologies in the power system will depend not only on their technical capabilities, but also on other externalities, such as those related to the regulatory framework and economic aspects, among others.

This chapter has also included an example on the utilization of secondary batteries for isolated power systems with PV generation. The example has addressed the dimensioning of a battery-based storage facility. In particular, the proposed sizing methodology yields the number of battery cells in series and in parallel needed so as to achieve the required energy capability, also taking into account the voltage and current limitations for utilization of the technology. The methodology also allows us to determine the type and number of power inverters for the battery bank.

References

1. Collier, S. (2014) *The Nexus of the Smart Grid and the Internet of Things*, IEEE Webinar, New York, http://smartgrid.ieee.org/resources/webinars/webinar-overviews/814-the-nexus-of-the-smart-grid-and-the-internet-of-things-with-steve-collier2 (accessed January 5, 2016).
2. Frost & Sullivan (2011) *The Changing Regulatory Landscape for Smart Energy. Impact on Information and Communication Technologies (ICT) Opportunities in Utilities*, Report, Frost & Sullivan, New York.
3. Frost & Sullivan (2011) *European Large Scale Energy Storage Market and Opportunities from Growth in Renewable Energy*, Report, Frost & Sullivan, New York.
4. European Commission (2006) *European Smart Grids Technology Platform: Vision and Strategy for Europe's Electricity*, http://ec.europa.eu/research/energy/pdf/smartgrids_en.pdf (accessed May 28, 2015).
5. Department of Energy and Climate Change, UK (2009) *The Smarter Grids: the Opportunity*, https://www.gov.uk/government/organisations/department-of-energy-climate-change (accessed May 28, 2015).
6. Toffler, A. (1980) *The Third Wave*, Bantam Books, New York.
7. Kundur, P. (1993) *Power System Stability and Control*, McGraw-Hill, New York.
8. Van der Sluis, L. (2001) *Transients in Power Systems*, John Wiley & Sons, Ltd, Chichester.
9. Handschin, E., Otero, A.F., and Cidrás, J. (2009) Steady-state single-phase models of power system components, in *Electric Energy Systems: Analysis and Operation* (eds A. Gómez-Expósito, A.J. Conejo, and C. Cañizares), CRC Press, Boca Raton, FL, pp. 55–94.
10. Gómez-Expósito, A., and Alvarado, F.L. (2009) Load flow*Electric Energy Systems: Analysis and Operation* (eds A. Gómez-Expósito, A.J. Conejo, and C. Cañizares), CRC Press, Boca Raton, FL, pp. 95–126.
11. Wood, A.J., Wollenberg, B.F., and Sheblé, G.B. (2014) *Power Generation, Operation and Control*, John Wiley & Sons, Ltd, Chichester.
12. Kirschen, D. and Strbac, G. (2004) *Fundamentals of Power System Economics*, John Wiley & Sons, Ltd, Chichester.
13. Gill, P. (2008) *Electrical Power Equipment Maintenance and Testing*, CRC Press, Boca Raton, FL.
14. Seifi, H. and Sepasian, M.S. (2011) *Electric Power System Planning: Issues, Algorithms and Solutions*, Springer-Verlag, Berlin.
15. Van Hertem, D., Gomis-Bellmunt, O., and Liang, J. (2015) *HVDC Grids for Transmission of Electrical Energy: Offshore Grids and a Future Supergrid & HVDC Transmission*, John Wiley & Sons, Ltd, Chichester.
16. ABB (1997) *Electrical Transmission and Distribution Reference Book*, ABB, Zurich.

Energy Storage in Power Systems, First Edition. Francisco Díaz-González, Andreas Sumper and Oriol Gomis-Bellmunt.
© 2016 John Wiley & Sons, Ltd. Published 2016 by John Wiley & Sons, Ltd.

17. ABB (2014) *Switchgear Manual*, ABB, Zurich.
18. Aragüés, M., Rimez, J., Beerten, J., *et al.* (2015) Secure and optimal operation of hybrid AC/DC grids with large penetration of offshore wind, in *ACDC 2015: AC and DC Power Transmission*. 11th International Conference on AC and DC Power Transmission, February 10–12, 2015, Edgbaston Stadium, Birmingham, UK. Institution of Engineering and Technology, Stevenage, Hertfordshire.
19. Gómez-Expósito, A., Conejo, A.J., and Cañizares, C. (eds) (2009) *Electric Energy Systems: Analysis and Operation*, CRC Press, Boca Raton, FL.
20. IRENA (2015) *Renewable Energy Statistics*, http://www.irena.org/home/ (accessed May 28, 2015).
21. World Energy Council (2013) *World Energy Perspective: Cost of Energy Technologies*, http://www.worldenergy.org/publications/2013/world-energy-perspective-cost-of-energy-technologies/ (accessed May 28, 2015).
22. Hau, E. (2005) *Wind Turbines: Fundamentals, Technologies, Application, Economics*, Springer-Verlag, Berlin.
23. Global Wind Energy Council (2014) *Global Wind Report: 2013 Annual Market Update*, http://www.gwec.net/publications/global-wind-report-2/ (accessed May 28, 2015).
24. EWEA (2015) *Wind in Power. 2014 European Statistics*, http://www.ewea.org/ (accessed May 28, 2015).
25. Tesla, T. (1901) *Apparatus for the Utilization of Radiant Energy*. US Patent 685,957.
26. Li, H. and Chen, Z. (2008) Overview of different wind generator systems and their comparisons. *IET Renewable Power Generation*, **2** (2), 123–138.
27. Slootweg, J., Polinder, H., and Kling, W.L. (2005) Reduced-order modelling of wind turbines, in *Wind Power in Power Systems* (ed. T. Ackermann), John Wiley & Sons, Ltd, Chichester, pp. 555–585.
28. Hansen, L.H., Helle, L., Blaabjerg, F., *et al.* (2001) *Conceptual Survey of Generators and Power Electronics for Wind Turbines*, Risø National Laboratory, Roskilde, Denmark.
29. Khadraoui, M.R. and Elleuch, M. (2008) Comparison between optislip and fixed speed wind energy conversion systems, in *5th International Multi-Conference on Systems, Signals and Devices*, July 20–22, 2008, Amman, Jordan. IEEE, Piscataway, NJ, pp. 1–6.
30. Peña, R., Clare, J.C., and Asher, G.M. (1996) Doubly fed induction generator using back-to-back PWM converters and its application to variable-speed wind-energy generation. *IEE Proceedings – Electric Power Applications*, **143** (3), 231–241.
31. Gomis-Bellmunt, O., Junyent-Ferre, A., Sumper, A., and Bergas-Jane, J. (2008) Ride-through control of a doubly fed induction generator under unbalanced voltage sags. *IEEE Transactions on Energy Conversion*, **23**, 1036–1045.
32. Hu, J. and Zhu, Z.Q. (2012) Electrical machines and power electronic systems for high-power wind energy penetration applications: part 1 – market penetration, current technology and advanced machine systems. *COMPEL: The International Journal for Computation and Mathematics in Electrical and Electronic Engineering*, **32** (4), 7–33.
33. Polinder, H., de Haan, S.W.H., Dubois, M.R., and Slootweg, J.G. (2005) Basic operation principles and electrical conversion systems of wind turbines, in *IEEE International Conference on Electric Machines and Drives*, May 15–15, 2005, San Antonio, TX. IEEE Press, Piscataway, NJ, pp. 543–550.
34. Dubois, M., and Polinder, H., and Ferreira, J., (2010) Comparison of generator topologies for direct-drive wind turbines. *Proceedings of Nordic Countries Power and Industrial Electronics Conference (NORPIE)*, Aalborg, pp. 22–26.
35. Polinder, H., Van der Pijl, F.F.A., de Vilder, G.-J., and Tavner, P. (2005) Comparison of direct-drive and geared generator concepts for wind turbines, in *IEEE International Conference on Electric Machines and Drives*, May 15–15, 2005, San Antonio, TX. IEEE Press, Piscataway, NJ, pp. 725–733.

36. El-Refaie, A.M. and Jahns, T.M. (2005) Optimal flux weakening in surface pm machines using fractional-slot concentrated windings. *IEEE Transactions on Industry Applications*, **41** (3), 790–800.
37. Walker, G. (2001) Evaluating MPPT converter topologies using a MATLAB PV model. *Journal of Electrical and Electronics Engineering*, Brisbane, **21** (1), 138–143.
38. Díaz-González, F., Hau, M., Sumper, A., and Gomis-Bellmunt, O. (2014) Participation of wind power plants in system frequency control: review of grid code requirements and control methods. *Renewable and Sustainable Energy Reviews*, **34**, 551–564.
39. Kayikci, M. and Milanovic, J. (2009) Dynamic contribution of DFIG-based wind plants to system frequency disturbances. *IEEE Transactions on Power Systems*, **24**, 859–867.
40. Yingcheng, X. and Nengling, T. (2011) Review of contribution to frequency control through variable speed wind turbine. *Renewable Energy*, **11**, 1671–1677.
41. Lalor, G., Mullane, A., and O'Malley, M. (2005) Frequency control and wind turbine technologies. *IEEE Transactions on Power Systems*, **20**, 1905–1913.
42. Díaz-González, F., Sumper, A., Gomis-Bellmunt, O., and Villafáfila-Robles, R. (2012) A review of energy storage technologies for wind power applications. *Renewable and Sustainable Energy Reviews*, **16**, 2154–2171.
43. EirGrid and System Operator for Northern Ireland (SONI) (2010). *All Island TSO Facilitation of Renewables Studies*, http://www.eirgrid.com (accessed April 26, 2014).
44. ENTSO-E (2013) *ENTSO-E Network Code for Requirements for Grid Connection Applicable to All Generators*, European Network of Transmission System Operators for Electricity, Brussels.
45. EirGrid (2013) *Eirgrid Grid Code Version 4.0*, http://www.eirgrid.com/operations/gridcode/ (accessed May 28, 2015).
46. Ackermann, T. (2005) *Wind Power in Power Systems*, John Wiley & Sons, Ltd, Chichester.
47. Sumper, A., Gomis-Bellmunt, O., Sudria-Andreu, A., *et al.* (2009) Response of fixed speed wind turbines to system frequency disturbances. *IEEE Transactions on Power Systems*, **24**, 181–192.
48. ENTSO-E (2009) *Operational Handbook. Policies. Load-Frequency Control and Performance*, European Network of Transmission System Operators for Electricity, Brussels.
49. ENTSO-E (2012) *Operational Reserve ad hoc Team Report: Final Version*, https://www.entsoe.eu/ (accessed April 26, 2014).
50. Schwab, A. (2009) *Elektroenergiesysteme, 2. Auflage*, Springer-Verlag, Berlin.
51. Sun, Y.-Z., Zhang, Z.-S., Li, G.-J., and Lin, J. (2010) Review on frequency control of power systems with wind power penetration, in *IEEE International Conference on Power System Technology (POWERCON)*, October 24–28, 2010, Hangzhou, China. IEEE Press, Piscataway, NJ, pp. 1–8.
52. Verband der Netzbetreiber – VDN – e.V. beim VDEW (2007) *Transmissioncode 2007: Network and System Rules of the German Transmission System Operators*, http://www.vde.com (accessed May 28, 2015).
53. Red Eléctrica de España (1998) *P.O 7.1 Servicio Complementario de Regulación primaria. Resolución de 30/7/1998, boe 18/08/98*, http://www.ree.es (accessed May 28, 2015).
54. Red Eléctrica de España (2009) *P.O. 7.2 Regulación Secundaria. Resolución de 18/5/2009, boe 28/05/09*, http://www.ree.es (accessed May 28, 2015).
55. National Grid plc (2012) *The Grid Code, Issue 4 Revision 13*, http://www.nationalgrid.com/uk/ (accessed May 28, 2015).
56. First Hydro Company, http://www.fhc.co.uk/pumped_storage.htm (accessed June 9, 2014).
57. KPMG (2010) *Central and Eastern European Hydro Power Outlook*, https://www.kpmg.com (accessed May 28, 2015).
58. Eurelectric (2011) *20% Renewables by 2020: a Eurelectric Action Plan, RESAP*, http://www.eurelectric.org/ (accessed May 28, 2015).

59. Erlich, I., Winter, W., and Dittrich, A. (2006) Advanced grid requirements for the integration of wind turbines into the German transmission system, in *IEEE Power Engineering Society General Meeting*, June 21, 2006, Montreal, Canada. IEEE Press, Piscataway, NJ, pp. 1–7.

60. Brisebois, J. and Aubut, N. (2011) Wind farm inertia emulation to fulfill Hydro-Québec's specific need, in *IEEE Power and Energy Society General Meeting*, July 24–29, 2011, San Diego, CA. IEEE Press, Piscataway, NJ, pp. 1–7.

61. Gonzalez-Longatt, F. (2012) Impact of synthetic inertia from wind power on the protection/control schemes of future power systems: simulation study, in *11th IET International Conference on Developments in Power Systems Protection (DPSP 2012)*, IET, Birmingham, UK. IEEE Press, Piscataway, NJ, pp. 1–6.

62. ENTSO-E (2012) *Network Code for Requirements for Grid Connection Applicable to all Generators: Frequently Asked Questions*, https://www.entsoe.eu/ (accessed April 15, 2014).

63. Junyent-Ferré, A., Gomis-Bellmunt, O., Sumper, A., *et al.* (2010) Modeling and control of the doubly fed induction generator wind turbine. *Simulation Modeling Practice and Theory*, **18**, 1365–1381.

64. Slootweg, J., Polinder, H., and Kling, W. (2003) Representing wind turbine electrical generating systems in fundamental frequency simulations. *IEEE Transactions on Energy Conversion*, **18**, 516–525.

65. El-Tous, Y. (2008) Pitch angle control of variable speed wind turbine. *American Journal of Engineering and Applied Sciences*, **1**, 118–120.

66. Bianchi, F.D., De Batista, H., and Mantz, R.J. (2007) *Wind Turbine Control Systems: Principles, Modeling and Gain Scheduling Design*, Springer, London.

67. Venne, P. and Guillaud, X. (2009) Impact of wind turbine controller strategy on deloaded operation, in *CIGRE/IEEE PES Joint Symposium Integration of Wide-Scale Renewable Resources Into the Power Delivery System*, July 29–31, 2009, Calgary, AB, Canada. IEEE Press, Piscataway, NJ.

68. Valsera-Naranjo, E., Sumper, A., Gomis-Bellmunt, O., *et al.* (2010) Pitch control system design to improve frequency response capability of fixed-speed wind turbine systems. *European Transactions on Electrical Power*, **21**, 1984–2006.

69. Camblong, H. (2008) Digital robust control of a variable speed pitch regulated wind turbine for above rated wind speeds. *Control Engineering Practice*, **16**, 946–958.

70. Camblong, H., Nourdine, S., Vechiu, I., and Tapia, G. (2012) Comparison of an island wind turbine collective and individual pitch LQG controllers designed to alleviate fatigue loads. *IET Renewable Power Generation*, **6**, 267–275.

71. Loukarakis, E., Margaris, I., and Moutis, P. (2009) Frequency control support and participation methods provided by wind generation, in *IEEE Electrical Power and Energy Conference (EPEC)*, October 22–23, 2009, Montreal, Canada. IEEE Press, Piscataway, NJ, pp. 1–6.

72. Janssens, N., Lambin, G., and Bragard, N. (2007) Active power control strategies of DFIG wind turbines, in *IEEE Lausanne Power Tech*, July 1–5, 2007, Lausanne, Switzerland. IEEE Press, Piscataway, NJ, pp. 516–521.

73. de Almeida, R., Castronuovo, E., and Peças-Lopes, J. (2006) Optimum generation control in wind parks when carrying out system operator requests. *IEEE Transactions on Power Systems*, **21**, 718–725.

74. Moutis, P., Papathanassiou, S., and Hatziargyriou, N. (2012) Improved load-frequency control contribution of variable speed variable pitch wind generators. *Renewable Energy*, **48**, 514–523.

75. Moutis, P., Loukarakis, E., Papathanasiou, S., and Hatziargyriou, N. (2009) Primary load-frequency control from pitch-controlled wind turbines, in *IEEE Power Tech*, June 28, 2009 – July 2, 2009, Bucharest, Romania. IEEE Press, Piscataway, NJ, pp. 1–7.

76. de Almeida, R. and Peças-Lopes, J. (2009) Participation of doubly fed induction wind generators in system frequency regulation. *IEEE Transactions on Power Systems*, **22**, 944–950.

77. Zertek, G., Verbic, A., and Pantos, M. (2012) Optimised control approach for frequency-control contribution of variable speed wind turbines. *IET Renewable Power Generation*, **6**, 17–23.

78. Zertek, G., Verbic, A., and Pantos, M. (2012) A novel strategy for variable-speed wind turbines' participation in primary frequency control. *IEEE Transactions on Sustainable Energy*, **3**, 791–799.

79. Ramtharan, J., Ekanayake, G., and Jenkins, N. (2007) Frequency support from doubly fed induction generator wind turbines. *IET Renewable Power Generation*, **1**, 3–9.

80. Zhang, Z.-S., Sun, Y., Lin, J., and Li, G.-J. (2012) Coordinated frequency regulation by doubly fed induction generation-based wind power plants. *IET Renewable Power Generation*, **6**, 38–47.

81. Rodriguez-Amenedo, J., Arnalte, S., and Burgos, J. (2002) Automatic generation control of a wind farm with variable speed wind turbines. *IEEE Transactions on Energy Conversion*, **17**, 279–284.

82. Abdelkafi, A. and Krichen, L. (2011) New strategy of pitch angle control for energy management of a wind farm. *Energy*, **36**, 1470–1479.

83. Holdsworth, L., Ekanayake, J., and Jenkins, N. (2004) Power system frequency response from a fixed speed and double fed induction generator-based wind turbines. *Wind Energy*, **7**, 21–35.

84. Haileselassie, T., Torres-Olguin, R., Vrana, T., and Uhlen, K. (2011) Main grid frequency support strategy for VSC–HVDC connected wind farms with variable speed wind turbines, in *IEEE Trondheim PowerTech*, June 19–23, 2011, Trondheim, Norway. IEEE Press, Piscataway, NJ, pp. 1–6.

85. Morren, J., Pierik, J., and de Haan, S. (2006) Inertial response of variable speed wind turbines. *Electric Power Systems Research*, **76**, 980–987.

86. Keung, P.-K., Li, P., Banakar, H., and Ooi, B. (2009) Kinetic energy of wind-turbine generators for system frequency support. *IEEE Transactions on Power Systems*, **24**, 279–287.

87. Conroy, J. and Watson, R. (2008) Frequency response capability of full converter wind turbine generators in comparison to conventional generation. *IEEE Transactions on Power Systems*, **23**, 649–656.

88. Mauricio, J., Marano, A., Gomez-Exposito, A., and Martinez-Ramos, J. (2009) Frequency regulation contribution through variable-speed wind energy conversion systems. *IEEE Transactions on Power Systems*, **24**, 173–180.

89. Pickard, W.F., Shen, Q.A., and Hansing, N.J. (2009) Parking the power: strategies and physical limitations for bulk energy storage in supply–demand matching on a grid whose input power is provided by intermittent sources. *Renewable and Sustainable Energy Reviews*, **13**, 1934–1945.

90. Ibrahim, H., Ilinca, A., and Perron, J. (2008) Energy storage systems – characteristics and comparisons. *Renewable and Sustainable Energy Reviews*, **12**, 1221–1230.

91. Dursun, B. and Alboyaci, B. (2010) The contribution of wind–hydro pumped storage systems in meeting Turkey's electric energy demand. *Renewable and Sustainable Energy Reviews*, **14**, 1979–1988.

92. Deane, J.P., Ó Gallachóir, B.P., and McKeogh, E.J. (2010) Techno-economic review of existing and new pumped hydro energy storage plant. *Renewable and Sustainable Energy Reviews*, **14**, 1293–1302.

93. Kaldellis, J.K. and Zafirakis, D. (2007) Optimum energy storage techniques for the improvement of renewable energy sources-based electricity generation economic efficiency. *Energy*, **32**, 2295–2305.

94. Zach, K., Auer, H., and Lettner, G. (2012) *stoRE Project: Current Status, Role and Costs of Energy Storage Technologies*, http://www.store-project.eu/ (accessed May 28, 2015).

95. Snihir, I., Rey, S., and Verbitskiy, E. (2010) *Battery Open-Circuit Voltage Estimation by a Method of Statistical Analysis*, http://www.eurandom.nl/reports/2005/046-report.pdf (accessed May 16, 2015).

96. IEEE Power & Energy Society (2010) *IEEE Recommended Practice for the Characterization and Evaluation of Emerging Energy Storage Technologies in Stationary Applications, IEEE Std 1679–2010*, IEEE, New York.
97. Ehsani, M., Gao, Y., and Emadi, A. (2014) *Modern Electric, Hybrid Electric, and Fuel Cell Vehicles: Fundamentals, Theory and Design*, CRC Press, Boca Raton, FL.
98. Broussely, M. and Pistoia, G. (eds) (2007) *Industrial Applications of Batteries: From Cars to Aerospace and Energy Storage*, Elsevier, Amsterdam.
99. Moriokaa, Y., Narukawab, S., and Itou, T. (2001) State-of-the-art of alkaline rechargeable batteries. *Journal of Power Sources*, **100**, 107–110.
100. NGK Insulators, Ltd, http://www.ngk.co.jp/english/ (accessed April 22, 2015).
101. Wen, Z., Cao, J., Gu, Z., *et al.* (2008) Research on sodium sulfur battery for energy storage. *Solid State Ionics*, **179**, 1697–1701.
102. Jalal Kazempour, S., Parsa Moghaddam, M., Haghifam, M.R., and Yousefi, G.R. (2009) Electric energy storage systems in a market-based economy: comparison of emerging and traditional technologies. *Renewable Energy*, **34**, 2630–2639.
103. Ter-Garzarian, A. (1994) *Energy Storage for Power Systems*, Peter Peregrinus, London.
104. Tesla Motors, http://www.teslamotors.com/powerwall (accessed May 29, 2015).
105. Wakihara, M. (2001) Recent developments in lithium ion batteries. *Materials Science and Engineering*, **33**, 109–126.
106. Bruce, P.G. (2008) Energy storage beyond the horizon: rechargeable lithium batteries. *Solid State Ionics*, **179**, 752–759.
107. Hadjipaschalis, I., Poullikkas, A., and Efthimiou, V. (2009) Overview of current and future energy storage technologies for electric power applications. *Renewable and Sustainable Energy Reviews*, **13**, 1513–1522.
108. IREC, Catalonia Institute for Energy Research, http://www.irec.cat/ (accessed April 22, 2015).
109. Hall, P.J. and Bain, E.J. (2008) Energy-storage technologies and electricity generation. *Energy Policy*, **36**, 4352–4354.
110. Beck, F. and Ruëtschi, P. (2000) Rechargeable batteries with aqueous electrolytes. *Electrochimica Acta*, **45**, 2467–2482.
111. Yang, J.H., Yang, H.S., Ra, H.W., *et al.* (2015) Effect of a surface active agent on performance of zinc/bromine redox flow batteries: improvement in current efficiency and system stability. *Journal of Power Sources*, **275**, 294–297.
112. Ponce-de-León, C., Frías-Ferrer, A., González-García, J., *et al.* (2006) Redox flow cells for energy conversion. *Journal of Power Sources*, **160**, 716–717.
113. Redflow Limited, http://www.redflow.com.au/ (accessed April 22, 2015).
114. DTI (2004). *Review of Electrical Energy Storage Technologies and Systems and of Their Potential for the UK*. Report, DTI, London.
115. Yogi-Goswami, D. and Kreith, F. (2007) *Energy Conversion*, CRC Press, Boca Raton, FL.
116. Ton, D.T., Hanley, C.J., Peek, G.H., and Boyes, J.D. (2008) *Solar Energy Grid Integration Systems – Energy Storage (SEGIS–ES)*, http://prod.sandia.gov/techlib/access-control.cgi/2008/084247.pdf (accessed April 22, 2015).
117. Smith, W. (2000) The role of fuel cells in energy storage. *Journal of Power Sources*, **86**, 74–83.
118. Prudent Energy Corporation, http://www.pdenergy.com/ (accessed April 22, 2015).
119. Winter, C-J. (2009) Hydrogen energy – abundant, efficient, clean: a debate over the energy-system-of-change. *International Journal of Hydrogen Energy*, **34**, 1–52.
120. Harrison, K.W., Remick, R., Martin, G.D., and Hoskin, A. (2010) Hydrogen production: fundamentals and case study summaries, in *Parallel Sessions Book 3: Hydrogen Production Technologies—Part 2* (eds D. Stolten and T. Grube). Proceedings of the 18th World Hydrogen Energy Conference, WHEC 2010, May 16–21, 2010, Essen, Germany. Forschungszentrum Jülich GmbH, Zentralbibliothek, Verlag, p. 3.

121. Conte, M., Iacobazzi, A., Ronchetti, M., and Vellone, R. (2001) Hydrogen economy for a sustainable development: state-of-the-art and technological perspectives. *Journal of Power Sources*, **100**, 171–187.

122. Li, P. (2008) Energy storage is the core of renewable energy technologies. *IEEE Nanotechnology Magazine*, **2**, 13–18.

123. Bolund, B., Bernhoff, H., and Leijon, M. (2007) Flywheel energy and power storage systems. *Renewable & Sustainable Energy Reviews*, **11**, 235–258.

124. ESA, Electricity Storage Association, http://www.electricitystorage.org/ (accessed May 7, 2015).

125. Lund, H. and Salgi, G. (2009) The role of compressed air energy storage (CAES) in future sustainable energy systems. *Energy Conversion and Management*, **50**, 1172–1178.

126. Denholm, P. and Kulcinski, G.L. (2004) Life cycle energy requirements and greenhouse gas emissions from large scale energy storage systems. *Energy Conversion and Management*, **45**, 2153–2172.

127. Barote, L., Weissbach, R., Teodorescu, *et al.* (2008) Stand-alone wind system with vanadium redox battery energy storage, in *11th International Conference on Optimization of Electrical and Electronic Equipment: OPTIM 2008*, May 22–24, 2008, Brasov, Romania. IEEE Press, Piscataway, NJ, pp. 407–412.

128. Huang, K.-L., Li, X.-G., Liu, S.-Q., *et al.* (2008) Research progress of vanadium redox flow battery for energy storage in China. *Renewable Energy*, **33**, 186–192.

129. Rydh, C.J. and Sandén, B.A. (2005) Energy analysis of batteries in photovoltaic systems, part II: energy return factors and overall battery efficiencies. *Energy Conversion and Management*, **46**, 1980–2000.

130. Parcon (2015) *Emerging Electric Technologies Whitepaper*, http://www.parcon.uci.edu/ (accessed May 7, 2015).

131. Esmaili, A. and Nasiri, A. (2010) Energy storage for short-term and long-term wind energy support, in *IECON 2010 – 36th Annual Conference on IEEE Industrial Electronics Society*, November 7–10, 2010, Glendale, AZ. IEEE Press, Piscataway, NJ, pp. 3281–3286.

132. Yamamura, T.T., Wu, X., Ohta, S., *et al.* (2011) Vanadium solid-salt battery: solid state with two redox couples. *Journal of Power Sources*, **196**, 4003–4009.

133. Baker, J. (2008) New technology and possible advances in energy storage. *Energy Policy*, **36**, 4368–4373.

134. Divya, K.C. and Ostergaard, J. (2009) Battery energy storage technology for power systems – an overview. *Electric Power Systems Research*, **79**, 511–520.

135. Dell, R.M. and Rand, D.A.J. (2001) Energy storage – a key technology for global energy sustainability. *Journal of Power Sources*, **100**, 2–17.

136. Rydh, C.J. (1999) Environmental assessment of vanadium redox and lead–acid batteries for stationary energy storage. *Journal of Power Sources*, **80**, 21–29.

137. Harding Energy, Inc., http://www.hardingenergy.com/ (accessed May 7, 2015).

138. McDowall, J. (2006) Integrating energy storage with wind power in weak electricity grids. *Journal of Power Sources*, **162**, 959–964.

139. Du-Pasquier, A., Plitz, I., Menocal, S., and Amatucci, G. (2003) A comparative study of Li-ion battery, supercapacitor and nonaqueous asymmetric hybrid devices for automotive applications. *Journal of Power Sources*, **115**, 171–178.

140. Adachi, K., Tajima, H., and Hashimoto, T. (2003) Development of 16 kWh power storage system applying Li-ion batteries. *Journal of Power Sources*, **119–121**, 897–901.

141. Ribeiro, P.F., Johnson, B.K., Crow, M.L., *et al.* (2001) Energy storage systems for advanced power applications. *Proceedings of the IEEE*, **89**, 1744–1766.

142. Ries, G. and Neumueller, H.-W. (2001) Comparison of energy storage in flywheels and SMES. *Physica C*, **357–360**, 1306–1310.

143. Liu, H. and Jiang, J. (2007) Flywheel energy storage – an upswing technology for energy sustainability. *Energy and Buildings*, **39**, 599–604.
144. Smith, S.C. and Sen, P.K. (2008) Ultracapacitors and energy storage: applications in electrical power system. In: *40th North American Power Symposium: NAPS '08*, September 28–30, 2008, Calgary, AB. IEEE Press, Piscataway, NJ, pp. 1–6.
145. Rafik, F., Gualous, H., Gallay, R., *et al.* (2007) Frequency, thermal and voltage supercapacitor characterization and modeling. *Journal of Power Sources*, **165**, 928–934.
146. Liu, Q., Nayfeh, N.H., and Yau, S.-T. (2010) Supercapacitor electrodes based on polyaniline–silicon nanoparticle composite. *Journal of Power Sources*, **195**, 3956–3959.
147. Cericola, D., Ruch, P.W., Kötz, R., *et al.* (2010) Simulation of a supercapacitor/Li-ion battery hybrid for pulsed applications. *Journal of Power Sources*, **195**, 2731–2736.
148. Maxwell Technologies, Inc., http://www.maxwell.com/ (accessed May 7, 2015).
149. MWH Global, Inc., http://www.mwhglobal.com/ (accessed May 7, 2015).
150. Benitez, L.E., Benitez, P.C., and Van Kooten, G.C. (2008) The economics of wind power with energy storage. *Energy Economics*, **30**, 1973–1989.
151. Fuel Cell Energy, Inc., http://www.fuelcellenergy.com/ (accessed May 7, 2015).
152. Dresser-Rand, http://www.dresser-rand.com/ (accessed May 7, 2015).
153. Vionx Energy, http://vionxenergy.com/products/ (accessed May 7, 2015).
154. Kondoh, J., Ishii, I., Yamaguchi, H., *et al.* (2000) Electrical energy storage systems for energy networks. *Energy Conversion & Management*, **41**, 1863–1874.
155. ZBB Energy Corporation, http://www.zbbenergy.com/ (accessed May 7, 2015).
156. Bito, A. (2005) Overview of the sodium–sulfur battery for the IEEE Stationary Battery Committee, in *IEEE Power Engineering Society General Meeting*, June 12–16, 2005, San Francisco, CA. IEEE Press, Piscataway, NJ, pp. 1232–1235.
157. Aifantis, K.E., Hackney, S.A., and Vasant-Kumar, R. (eds) (2010) *High Energy Density Lithium Batteries*, Wiley-VCH Verlag GmbH, Weinheim.
158. Alcad, http://www.alcad.com/ (accessed May 7, 2015).
159. Exide Technologies, http://www.exide.com/ (accessed May 7, 2015).
160. Barton, J.P. and Infield, D.G. (2004) Energy storage and its use with intermittent renewable energy. *IEEE Transactions on Energy Conversion*, **19**, 441–448.
161. Saft Batteries, http://www.saftbatteries.com/ (accessed May 7, 2015).
162. Clark, N.H. and Doughty, D.H. (2005) Development and testing of 100 kW/1 min Li-ion battery systems for energy storage applications. *Journal of Power Sources*, **146**, 798–803.
163. A123 Systems, http://www.a123systems.com/ (accessed May 7, 2015).
164. Li-Tec Battery GmbH, http://www.li-tec.de/en/ (accessed May 7, 2015).
165. Ali, M.H., Wu, B., and Dougal, R.A. (2010) An overview of SMES applications in power and energy systems. *IEEE Transactions on Sustainable Energy*, **1**, 38–47.
166. Beaudin, M., Zareipour, H., Schellenberglabe, A., and Rosehart, W. (2010) Energy storage for mitigating the variability of renewable electricity sources: an updated review. *Energy for Sustainable Development*, **14**, 302–314.
167. Superconductor Technologies, Inc., http://www.suptech.com/ (accessed May 7, 2015).
168. SuperPower, Inc. (2010) *Superconducting Magnetic Energy Storage (SMES) Project*, http://www.superpower-inc.com/content/superconducting-magnetic-energy-storage-smes (accessed April 22, 2015).
169. Beacon Power Corporation, http://www.beaconpower.com (accessed May 7, 2015).
170. Active Power, http://www.activepower.com/ (accessed May 7, 2015).
171. Piller Power Systems, http://www.piller.com (accessed May 7, 2015).
172. Barker, P.P. (2002) Ultracapacitors for use in power quality and distributed resource applications, in *IEEE Power Engineering Society Summer Meeting*, July 21–25, 2002, Chicago, IL. IEEE Press, Piscataway, NJ, pp. 316–320.

173. Kotz, R. and Carlen, M. (2000) Principles and applications of electrochemical capacitors. *Electrochimica Acta*, **45**, 2483–2498.
174. EPCOS AG, http://www.epcos.com/ (accessed May 7, 2015).
175. NEC TOKIN Corporation, http://www.nec-tokin.com/english/ (accessed May 7, 2015).
176. Nielsen, K.E. (2010) Superconducting magnetic energy storage in power systems with renewable energy sources. Master thesis. Norwegian University of Science and Technology.
177. Gualous, H., Bouquain, D., Berthon, A., and Kauffmann, J.M. (2003) Experimental study of supercapacitor serial resistance and capacitance variations with temperature. *Journal of Power Sources*, **123**, 86–93.
178. Sharma, P. and Bhatti, T.S. (2010) A review on electrochemical double-layer capacitors. *Energy Conversion and Management*, **51**, 2901–2912.
179. Payman, A., Pierfederici, S., and Meibody-Tabar, F. (2008) Energy control of supercapacitor/fuel cell hybrid power source. *Energy Conversion and Management*, **49**, 1637–1644.
180. Gil, A., Arce, P., Martorell, I., *et al.* (2014) *State of the Art of High Temperature Storage in Thermosolar Plants*, Universitat de Lleida, Lleida.
181. Kuravi, S., Trahan, J., Yogi-Goswami, D., *et al.* (2013) Thermal energy storage technologies and systems for concentrating solar power plants. *Progress in Energy and Combustion Science*, **39**, 285–319.
182. Pomianowski, M., Heiselberg, P., and Zhang, Y. (2013) Review of thermal energy storage technologies based on PCM application in buildings. *Energy and Buildings*, **67**, 56–69.
183. Solar Millennium AG (2015) *The Parabolic Trough Power Plants Andasol 1 to 3*, http://www.rwe.com/ (accessed May 28, 2015).
184. BMWI (2014) *Power-to-Gas (PtG) in Transport Status Quo and Perspectives for Development*, Federal Ministry of Transport and Digital Infrastructure, http://www.lbst.de/ (accessed May 28, 2015).
185. Newton, J. (2014) Power-to-gas and methanation – pathways to a hydrogen economy. Presented at 14th Annual APGTF Workshop, March 12–13, 2014, London.
186. Grond, L., Schulze, P., and Holstein, J. (2013) *Part of TKI Project TKIG01038 – Systems Analyses Power-to-Gas Pathways. Deliverable 1: Technology Review*, http://www.northseapowertogas.com/documents (accessed May 28, 2015).
187. Deutsche Energie-Agentur GmbH, http://www.dena.de/en/about-dena.html (accessed March 20, 2015).
188. North Sea Power to Gas Platform, http://www.northseapowertogas.com/about/power-to-gas (accessed March 20, 2015).
189. Pires, F.L., Antunes, F., and Oliveira, D. Jr. (2010) A neutral point clamped multilevel rectifier for grid connected wind energy systems, in *XVIII Congresso Brasileiro de Automática*, Bonito, pp. 1583–1587.
190. Luo, F.L. and Ye, H. (2010) *Power Electronics: Advanced Conversion Technologies*, CRC Press, Boca Raton, FL.
191. Barker, C.D., Davidson, C.C., Trainer, D.R., and Whitehouse, R.S. (2012) Requirements of DC–DC converters to facilitate large DC grids. *CIGRE Report B4-204.*
192. Siemens AG (2015) *Siestorage, The Modular Energy Storage System for a Sustainable Energy Supply*, http://www.siemens.com/siestorage (accessed May 28, 2015).
193. Alstom (2015) *Alstom's "MaxSine^{TM} eStorage"*, http://www.alstom.com/microsites/grid/products-and-services/battery-energy-storage/ (accessed April 25, 2015).
194. Alstom (2015) *Nice Grid Project: Smart Grids Shape the City of Tomorrow*, http://www.alstom.com/press-centre/2014/2/nice-grid-project-smart-grids-shape-the-city-of-tomorrow/ (accessed April 25, 2015).
195. ABB, http://www.abb.com/ (accessed April 25, 2015).

196. Wade, N., Taylor, P., Lang, P., and Svensson, J. (2009) Energy storage for power flow management and voltage control on an 11 kV UK distribution network, in *CIRED 2009: The 20th International Conference on Electricity Distribution*, June 8–11, 2009, Prague, Czech Republic. IEEE Press, Piscataway, NJ, pp. 1–4.

197. Schaefer, Inc. (2015) *Thyristor-controlled Power Supplies and Battery Chargers*, http://www.schaeferpower.com (accessed April 25, 2015).

198. SMA Solar Technology AG, http://www.sma.de/en.html (accessed April 25, 2015).

199. SMA Solar Technology AG (2015) *Battery Chargers for TGV POS Traction Units*, http://www.sma-railway.com/ (accessed April 25, 2015).

200. Davis, A., Salameh, Z.M., and Eaves, S.S. (1999) Evaluation of lithium-ion synergetic battery pack as battery charger. *IEEE Transactions on Energy Conversion*, **14** (3), 830–835.

201. ADC Krone (2008) *TIA-942 Data Centre Standards Overview*, http://www.herts.ac.uk/ (accessed April 26, 2015).

202. Zigor Corporación SA, http://www.zigor.com/eu/index.php?lang=en (accessed April 26, 2015).

203. Jossen, A. Spath, V., Doring, H., and Garche, J. (1999) Battery Management Systems (BMS) for increasing battery life time, in *The Third International Telecommunications Energy Special Conference*, Dresden, pp. 81–88.

204. Lauder, M.T., Suthar, B., and Northrop, P.W.C. (2014) Battery Energy Storage System (BESS) and Battery Management System (BMS) for grid-scale applications. *Proceedings of the IEEE*, **102** (6), 1014–1030.

205. Zakeri, B. and Syri, S. (2015) Electrical energy storage systems: a comparative life cycle cost analysis. *Renewable and Sustainable Energy Reviews*, **42**, 569–596.

206. Schoenung, S.M. and Hassenzahl, W.V. (2003) Long- vs. Short-term Energy Storage Technologies Analysis. *Sandia National Laboratories Report SAND2003-2783*, http://prod.sandia.gov/techlib/access-control.cgi/2003/032783.pdf (accessed March 24, 2015).

207. Power Sonic Corporation (2015) *Sealed Lead–Acid Batteries, Technical Manual*, http://www.power-sonic.com/ (accessed March 29, 2015).

208. National Renewable Energy Laboratory (NREL) (2010) *Hydrogen for Energy Storage Analysis Overview*, http://www.nrel.gov/hydrogen/pdfs/48360.pdf (accessed May 28, 2015).

209. National Renewable Energy Laboratory (NREL) (2009) *Current State-of-the-Art Hydrogen Production Cost Estimate Using Water Electrolysis*, http://www.hydrogen.energy.gov/pdfs/46676.pdf (accessed May 28, 2015).

210. Maxwell Technologies, Inc. (2015). *White Paper: Ultracapacitor Applications for Uninterruptible Power Supplies (UPS)*, http://www.maxwell.com/images/documents/whitepaper_application_for_ups.pdf (accessed May 28, 2015).

211. OPAL-RT Technologies, http://www.opal-rt.com/index (accessed May 29, 2015).

212. Díaz-González, F., Sumper, A., Gomis-Bellmunt, O., and Bianchi, F.D. (2013) Energy management of flywheel-based energy storage device for wind power smoothing. *Applied Energy*, **110**, 207–219.

213. Krause, P.C., Wasynczuk, O. and Sudhoff, S.D. (2002) *Analysis of Electric Machinery and Drive Systems*, John Wiley & Sons, Inc., Hoboken, NJ.

214. Akagi, H., Watanabe, E.H., and Aredes, M. (2007) *Instantaneous Power Theory and Applications to Power Conditioning*, John Wiley & Sons, Inc., Hoboken, NJ.

215. Domínguez-García, J.L., Gomis-Bellmunt, O., Trilla-Romero, L., and Junyent-Ferré, A. (2011) Indirect vector control of a squirrel cage induction generator wind turbine. *Computers & Mathematics with Applications*, **64**, 102–114.

216. Gomis-Bellmunt, O., Junyent-Ferré, A., Sumper, A., and Bergas-Jane, J. (2010) Permanent magnet synchronous generator offshore wind farms connected to a single power converter, in *IEEE Power and Energy Society General Meeting*, July 25–29, 2010, Minneapolis, MN. IEEE Press, Piscataway, NJ, pp. 1–6.

217. Aström, K.J. and Murray, R.M. (2008) *Feedback Systems: An Introduction for Scientists and Engineers*, Princeton University Press, Princeton, NJ.
218. Holmes, G. and Lipo, T.A. (2003) *Pulse Width Modulation for Power Converters: Principles and Practice*, John Wiley & Sons, Inc., Hoboken, NJ.
219. Wu, B., Lang, Y., Zargari, N., and Kouro, S. (2011) *Power Conversion and Control of Wind Energy Systems*, John Wiley & Sons, Hoboken, NJ.
220. Kazmierkowski, M.P., Krishnam, R., and Blaabjerg, F. (2002) *Control in Power Electronics, Selected Problems*. Academic Press, London.
221. Tsang, K.M. and Chan W.L. (2005) Cascade controller for DC/DC buck convertor. *IEEE Proceedings in Electrical Power Applications*, **152**, 827–831.
222. Kim, T. and Qiao W. (2011) A hybrid battery model capable of capturing dynamic circuit characteristics and nonlinear capacity effects. *IEEE Transactions on Energy Conversion*, **26**, 1172–1180.
223. Nielsen, K.E. and Molinas, M. (2010) Superconducting Magnetic Energy Storage (SMES) in power systems with renewable energy sources, in *IEEE International Symposium on Industrial Electronics (ISIE)*, July 4–7, 2010, Bari, Italy. IEEE Press, Piscataway, NJ, pp. 2487–2492.
224. Jongerden, M.R. and Haverkort, B.R. (2009) Which battery model to use? *IET Software*, **3**, 445–457.
225. Peukert, W. (1897) Über die Abhängigkeit der Kapazität von der Entladestromstärke bei Bleiakkumulatoren. *Elektrotechnische Zeitschrift*, **20**, pp. 20–21.
226. Rakhmatov, D., Vrudhula, S., and Wallach, A. (2001) An analytical high-level battery model for use in energy management of portable electronic systems, in *Proceedings of the 2001 IEEE/ACM International Conference on Computer-Aided Design (ICCAD 2001)*, November 4–8, 2001, San Jose, CA. IEEE Press, Piscataway, NJ, pp. 448–493.
227. Li, J., Fang, Q., Liu, F., and Liu, Y. (2014) Analytical modeling of dislocation effect on diffusion induced stress in a cylindrical lithium ion battery electrode. *Journal of Power Sources*, **272**, 121–127.
228. Manwell, J.F. and McGowan, J.G. (1993) Lead acid battery storage model for hybrid energy systems. *Solar Energy*, **50**, 399–405.
229. Bumby, J.R., Clarke, P.H., and Forster, I. (1985) Computer modelling of the automotive energy requirements for internal combustion engine and battery electric-powered vehicle. *IEEE Proceedings*, **132** (5), 265–279.
230. Williamson, S., Rimmalapudi, C., and Emadi, A. (2004) Electrical modeling of renewable energy sources and energy storage devices. *Journal of Power Electronics*, **4**, 117–121.
231. Rekioua, D. and Matagne, E. (2012) *Optimization of Photovoltaic Power Systems: Modelization, Simulation and Control*, Springer, London.
232. Appebaum, J. and Weiss, R. (1982) Estimation of battery charge in photovoltaic systems, in *16th IEEE Photovoltaic Specialists Conference*, June 14–17, 1982, New York. IEEE Press, Piscataway, NJ, pp. 513–518.
233. Copetti, J.B., Lorenzo, E., and Chenlo, F. (1993) A general battery model for PV systems simulation. *Progress in Photovoltaics: Research and Applications*, **1**, 283–292.
234. Shepherd, C.M. (1962) *Theoretical Design of Primary and Secondary Cells. Part III – Battery Discharge Equation*, US Naval Research Laboratory, Washington, DC.
235. Nasar, S.A. and Unnewehr, L.E. (1983) *Electromechanics and Electric Machines*, John Wiley & Sons, Inc., New York.
236. Tremblay, O. and Dessaint, L.-A. (2009) Experimental validation of a battery dynamic model for EV applications, in *EVS24 International Battery, Hybrid and Fuel Cell Electric Vehicle Symposium*, AVERE, Stavanger, pp. 1–9.
237. Salameh, Z.M., Casacca, M.A., and Lynch, W.A. (1992) A mathematical model for lead acid batteries. *IEEE Transactions on Energy Conversion*, **7**, 93–98.

238. Sidhu, A., Izadian, A., and Anwar, S. (2015) Adaptive nonlinear model-based fault diagnosis of Li-ion batteries. *IEEE Transactions on Industrial Electronics*, **62**, 1002–1011.
239. Buller, S., Thele, M., De Doncker, R.W.A.A., and Karden, E. (2005) Impedance-based simulation models of supercapacitors and Li-ion batteries for power electronic applications. *IEEE Transactions on Industry Applications*, **41** (3), 742–747.
240. Do, D.V., Forgez, C., Benkara, K.E.K., and Friedrich, G. (2009) Impedance observer for a Li-ion battery using Kalman filter. *IEEE Transactions on Vehicular Technology*, **58**, 3930–3937.
241. Randles, J.E.B. (1947) Kinetics of rapid electrode reactions. *Discussions of the Faraday Society*, **1**, 11–19.
242. Einhorn, M., Valerio Conte, F., Kral, C., and Fleig, J. (2013) Comparison, selection, and parameterization of electrical battery models for automotive applications. *IEEE Transactions on Power Electronics*, **28**, 1429–1437.
243. Dai, H., Zhang, X., Wei, X., *et al.* (2013) Cell-BMS validation with a hardware-in-the-loop simulation of lithium-ion battery cells for electric vehicles. *Electrical Power and Energy Systems*, **52**, 174–184.
244. Sritharan, T., Yan, D., Dawson, F.P., and Lian, K.K. (2014) Characterizing battery behavior for time-dependent currents. *IEEE Transactions on Industry Applications*, **50**, 4090–4097.
245. Unamuno, E., Gorrotxategi, L., and Aizpuru, I. (2014) Li-ion battery modeling optimization based on electrical impedance spectroscopy measurements, in *International Symposium on Power Electronics, Electrical Drives, Automation and Motion (SPEEDAM)*, June 18–20, 2014, Ischia, Italy. IEEE, Piscataway, NJ, pp. 154–160.
246. Terorde, G. (2004) *Electrical Drives and Control Techniques*, Acco, Leuven.
247. Stulrajter, M., Hrabovcová, V., and Franko, M. (2007) Permanent magnets synchronous motor control theory. *Journal of Electrical Engineering*, **58**, 79–84.
248. Cárdenas, R., Peña, R., Asher, G., *et al.* (2004) Control strategies for power smoothing using a flywheel driven by a sensorless vector-controlled induction machine operating in a wide speed range. *IEEE Transactions on Industrial Electronics*, **51**, 603–614.
249. Krishnan, R. (1993) Control and operation of PM synchronous motor drives in the field-weakening region, in *International Conference on Industrial Electronics, Control and Instrumentation: IECON '93*, November 15–19, 1993, Maui, HI. IEEE Press, Piscataway, NJ, pp. 745–750.
250. Díaz-González, F., Sumper, A., Gomis-Bellmunt, O., and Villafáfila-Robles, R. (2013) Modeling, control and experimental validation of a flywheel-based energy storage device. *Journal of European Power Electronics*, **23**, 1–21.
251. Bayod-Rújula, A.A. (2009) Future development of the electricity systems with distributed generation. *Energy*, **34**, 377–383.
252. EPRI (2004) *EPRI–DOE Handbook Supplement of Energy Storage for Grid Connected Wind Generation Applications*, Electric Power Research Institute, Concord, CA.
253. Georgilakis, P.S. (2008) Technical challenges associated with the integration of wind power into power systems. *Renewable and Sustainable Energy Reviews*, **12**, 852–863.
254. Świerzyński, M., Teodorescu, R., Rasmussen, C.N., *et al.* (2010) Storage possibilities for enabling higher wind energy penetration, in *EPE Wind Power Symposium*, April 15–16, 2010, Staffordshire University, Stafford, UK.
255. Ahmad-Lone, S.A. and Mufti, M.-u-D. (2006) Integrating a redox flow battery system with a wind–diesel power system, in *International Conference on Power Electronics, Drives and Energy Systems: PEDES '06*, December 12–15, 2006, New Delhi, India. IEEE Press, Piscataway, NJ, pp. 1–6.
256. Miyake, S. and Tokuda, N. (2001) Vanadium redox-flow battery for a variety of applications, in *IEEE Power Engineering Society Summer Meeting*, July 15–19, 2001, Vancouver, BC. IEEE Press, Piscataway, NJ, pp. 450–451.

257. Wang, W., Ge, B., Bi, D., and Sun, D. (2010) Grid-connected wind farm power control using VRB-based energy storage system, in *IEEE Energy Conversion Congress and Exposition (ECCE)*, September 12–16, 2010, Atlanta, GA. IEEE Press, Piscataway, NJ, pp. 3772–3777.

258. Yoshimoto, K., Nanahara, T., and Koshimizu, G. (2009) Analysis of data obtained in demonstration test about battery energy storage system to mitigate output fluctuation of wind farm, in *CIGRE/IEEE PES Joint Symposium Integration of Wide-Scale Renewable Resources Into the Power Delivery System*, July 29–31, 2009, Calgary, AB, Canada. IEEE Press, Piscataway, NJ.

259. Wang, W., Ge, B., Bi, D., *et al.* (2010) Energy storage based LVRT and stabilizing power control for direct-drive wind power system, in *International Conference on Power System Technology (POWERCON)*, October 24–28, 2010, Hangzhou, China. IEEE Press, Piscataway, NJ, pp. 1–6.

260. Van der Linden, S. (2006) Bulk energy storage potential in the USA, current developments and future prospects. *Energy*, **31**, 3446–3457.

261. Iqbal, M.T. (2003) Modeling and control of a wind fuel cell hybrid energy system. *Renewable Energy*, **28**, 223–237.

262. Baumann, L., Boggasch, E., Rylatt, M., and Wright, A. (2010) Energy flow management of a hybrid renewable energy system with hydrogen, in *IEEE Conference on Innovative Technologies for an Efficient and Reliable Electricity Supply (CITRES)*, September 27–29, 2010, Waltham, MA. IEEE Press, Piscataway, NJ, pp. 78–85.

263. Daneshi, H., Daneshi, A., Tabari, N.M., and Jahromi, A.N. (2009) Security-constrained unit commitment in a system with wind generation and compressed air energy storage, in *EEM 2009: 6th International Conference on the European Energy Market*, May 27–29, Leuven, Belgium. IEEE Press, Piscataway, NJ, pp. 1–6.

264. Sasaki, T., Kadoya, T., and Enomoto, K. (2004) Study on load frequency control using redox flow batteries. *IEEE Transactions on Power Systems*, **19**, 660–667.

265. Muljadi, E., Butterfield, C.P., Chacon, J., and Romanowitz, H. (2006) Power quality aspects in a wind power plant, in *IEEE Power Engineering Society General Meeting*, June 21, 2006, Montreal, Canada. IEEE Press, Piscataway, NJ, pp. 1–8.

266. Shi, G., Cao, Y.F., Li, Z., and Cai, X. (2010) Impact of wind-battery hybrid generation on isolated power system stability, in *International Symposium on Power Electronics, Electrical Drives, Automation and Motion (SPEEDAM)*, June 14–16, 2010, Pisa, Italy. IEEE Press, Piscataway, NJ, pp. 757–761.

267. Sebastián, R. (2011) Modelling and simulation of a high penetration wind diesel system with battery energy storage. *Electrical Power and Energy Systems*, **33**, 767–774.

268. Sebastián, R. and Peña-Alzola, R. (2010) Effective active power control of a high penetration wind diesel system with a Ni–Cd battery energy storage. *Renewable Energy*, **35**, 952–965.

269. Kuperman, A. and Aharon, I. (2011) Battery–ultracapacitor hybrids for pulsed current loads: a review. *Renewable and Sustainable Energy Reviews*, **15**, 981–992.

270. Hayashi, H., Hatabe, Y., Nagafuchi, T., *et al.* (2006) Test results of power system control by experimental SMES. *IEEE Transactions on Applied Superconductivity*, **16**, 598–601.

271. Ise, T., Kita, M., and Taguchi, A. (2005) A hybrid energy storage with a SMES and secondary battery. *IEEE Transactions on Applied Superconductivity*, **15**, 1915–1918.

272. Jung, H.Y., Kim, A.-R., Kim, J.-H., *et al.* (2009) A study on the operating characteristics of SMES for the dispersed power generation system. *IEEE Transactions on Applied Superconductivity*, **19**, 2028–2031.

273. Kim, A.-R., Seo, H.-R., Kim, G.H., *et al.* (2010) Operating characteristic analysis of HTS SMES for frequency stabilization of dispersed power generation system. *IEEE Transactions on Applied Superconductivity*, **20**, 1334–1338.

274. Kinjo, T., Senjyu, T., Urasaki, N., and Fujita, H. (2006) Terminal-voltage and output-power regulation of wind-turbine generator by series and parallel compensation using SMES. *IEE Proceedings – Generation, Transmission and Distribution*, **153**, 276–282.

275. Nomura, S., Ohata, Y., Hagita, T., *et al.* (2005) Wind farms linked by SMES systems. *IEEE Transactions on Applied Superconductivity*, **15**, 1951–1954.

276. Shi, J., Tang, Y.J., Ren, L., *et al.* (2008) Application of SMES in wind farm to improve voltage stability. *Physica C*, **468**, 2100–2103.

277. Boukettaya, G., Krichen, L., and Ouali, A. (2010) A comparative study of three different sensorless vector control strategies for a flywheel energy storage system. *Energy*, **35**, 132–139.

278. Ghedamsi, K., Aouzellag, D., and Berkouk, E.M. (2008) Control of wind generator associated to a flywheel energy storage system. *Renewable Energy*, **33**, 2145–2156.

279. Leclercq, L., Robyns, B., and Grave, J.-M. (2003) Control based on fuzzy logic of a flywheel energy storage system associated with wind and diesel generators. *Mathematics and Computers in Simulation*, **63**, 271–280.

280. Ran, L., Xiang, D., and Kirtley, J. (2011) Analysis of electromechanical interactions in a flywheel system with a doubly-fed induction machine. *IEEE Transactions on Industry Applications*, **47**, 1498–1506.

281. Ray, P.K., Mohanty, S.R., and Kishor, N. (2011) Proportional–integral controller based small-signal analysis of hybrid distributed generation systems. *Energy Conversion and Management*, **52**, 1943–1954.

282. Surive, G.O. and Mercado, P.E. (2010) DSTATCOM with flywheel energy storage system for wind energy applications: control design and simulation. *Electric Power Systems Research*, 80, 345–353.

283. Cárdenas, R., Peña, R., Asher, G., and Clare, J. (2001) Control strategies for enhanced power smoothing in wind energy systems using a flywheel driven by a vector-controlled induction machine. *IEEE Transactions on Industrial Electronics*, **48**, 625–635.

284. Cárdenas, R., Peña, R., Pérez, M., *et al.* (2006) Power smoothing using a flywheel driven by a switched reluctance machine. *IEEE Transactions on Industrial Electronics*, **53**, 1086–1093.

285. Abbey, C. and Joos, G. (2007) Supercapacitor energy storage for wind energy applications. *IEEE Transactions on Industry Applications*, **43**, 769–776.

286. Helwig, A. and Ahfock, T. (2009) Ultra-capacitor assisted battery storage for remote area power supplies: a case study, in *AUPEC 2009: 19th Australasian Universities Power Engineering Conference: Sustainable Energy Technologies and Systems*, September 27–30, 2009, Adelaide, SA. IEEE Press, Piscataway, NJ, pp. 1–6.

287. Qu, L. and Qiao, W. (2011) Constant power control of DFIG wind turbines with supercapacitor energy storage. *IEEE Transactions on Industry Applications*, **47**, 359–367.

288. Bhatia, R.S., Jain, S.P., Jain, D.K., and Singh, B. (2005) Battery energy storage system for power conditioning of renewable energy sources, in *International Conference on Power Electronics and Drives Systems: PEDS 2005*, November 28 – December 1, 2005, Kuala Lumpur, Malaysia. IEEE Press, Piscataway, NJ, pp. 501–506.

289. Wang, X.Y., Vilathgamuwa, D.M., and Choi, S.S. (2008) Buffer scheme with battery energy storage capability for enhancement of network transient stability and load ride-through. *Journal of Power Sources*, **179**, 819–829.

290. Ali, M.H., Park, M., Yu, I.-K. *et al.* (2009) Improvement of wind generator stability by fuzzy logic-controlled SMES. *IEEE Transactions on Industry Applications*, **45**, 1045–1051.

291. Arsoy, A.B., Liu, Y., Ribeiro, P.F., and Wang, F. (2003) Static-synchronous compensators and superconducting magnetic energy storage systems in controlling power system dynamics. *IEEE Industry Applications Magazine*, 21–28.

292. Kim, H.J., Seong, K.C., Cho, J.W., *et al.* (2006) 3 MJ/750 kVA SMES system for improving power quality. *IEEE Transactions on Applied Superconductivity*, **16**, 574–577.

293. Liu, F., Mei, S., Xia, D., *et al.* (2004) Experimental evaluation of nonlinear robust control for SMES to improve the transient stability of power systems. *IEEE Transactions on Energy Conversion*, **19**, 774–782.

294. Padimiti, D.S. and Chowdhury, B.H. (2007) Superconducting Magnetic Energy Storage System (SMES) for improved dynamic system performance, in *IEEE Power Engineering Society General Meeting*, June 24–28, 2007, Tampa, FL. IEEE Press, Piscataway, NJ, pp. 1–6.

295. Barrado, J.A., Griñó, R., and Valderrama-Blavi, H. (2010) Power-quality improvement of a stand-alone induction generator using a STATCOM with battery energy storage system. *IEEE Transactions on Power Delivery*, **25**, 2734–2741.

296. Parker, C.D. (2001) Lead–acid battery energy-storage systems for electricity supply networks. *Journal of Power Sources*, **100**, 18–28.

297. Tsai, M.-T., Lin, C.-E., Tsai, W.-I., and Huang, C.-L. (1995) Design and implementation of a demand-side multifunction battery energy storage system. *IEEE Transactions on Industrial Electronics*, **42**, 642–652.

298. Virulkar, V., Aware, M., and Kolhe, M. (2011) Integrated battery controller for distributed energy system. *Energy*, **36**, 2392–2398.

299. Wang, L., Chen, S.-S., Lee, W.-J., and Chen, Z. (2009) Dynamic stability enhancement and power flow control of a hybrid wind and marine-current farm using SMES. *IEEE Transactions on Energy Conversion*, **24**, 626–639.

300. Du, W., Wang, H.F., Cheng, S., *et al.* (2011) Robustness of damping control implemented by energy storage systems installed in power systems. *Electrical Power and Energy Systems*, **33**, 35–42.

301. Hirabayashi, S., Tomita, Y., and Iwamoto, S. (2008) Enhancement of transient stability ATC using NAS battery systems, in *Transmission and Distribution Conference and Exposition: 2008 IEEE/PES: Powering Toward the Future*, April 21–24, 2008, Chicago, IL. IEEE Press, Piscataway, NJ, pp. 1–6.

302. Ali, M.H. and Wu, B. (2010) Comparison of stabilization methods for fixed-speed wind generator systems. *IEEE Transactions on Power Delivery*, **25**, 323–331.

303. Muyeen, S.M., Hasan-Ali, M., Takahashi, R., *et al.* (2008) Damping of blade-shaft torsional oscillations of wind turbine generator system. *Electric Power Components and Systems*, **36**, 195–211.

304. Ngamroo, I., Cuk-Supriyadi, A.N., Dechanupaprittha, S., and Mitani, Y. (2009) Power oscillation suppression by robust SMES in power system with large wind power penetration. *Physica C*, **469**, 44–51.

305. Liu, D.B., Shi, L.J., Xu, Q., *et al.* (2009) Selection of installing locations of flywheel energy storage system in multimachine power systems by modal analysis, in *International Conference on Sustainable Power Generation and Supply: SUPERGEN '09*, April 6–7, 2009, Nanjing, China. IEEE Press, Piscataway, NJ, pp. 1–4.

306. Zhong, Y., Zhang, J., Li, G., and Chen, Z. (2006) Research on restraining low frequency oscillation with flywheel energy storage system, in *International Conference on Power System Technology: PowerCon 2006*, October 22–26, 2006, Chongqing, China. IEEE Press, Piscataway, NJ, pp. 1–4.

307. Mufti, M., Lone, S.A., Iqbal, S.J., *et al.* (2009) Super-capacitor based energy storage system for improved load frequency control. *Electric Power Systems Research*, **79**, 226–233.

308. Lee, D.-J. and Wang, L. (2008) Small-signal stability analysis of an autonomous hybrid renewable energy power generation/energy storage system part I: time-domain simulations. *IEEE Transactions on Energy Conversion*, **23**, 311–320.

309. Mercier, P., Cherkaoui, R., and Oudalov, A. (2009) Optimizing a battery energy storage system for frequency control application in an isolated power system. *IEEE Transactions on Power Systems*, **24**, 1469–1477.

310. Wagner, R. (1997) Large lead/acid batteries for frequency regulation, load levelling and solar power applications. *Journal of Power Sources*, **67**, 163–172.

311. AEA (2009) *Energy Storage Review*, Alaska Energy Authority, Alaska.

312. Lazarewicz, M.L. and Rojas, A. (2004) Grid frequency regulation by recycling electrical energy in flywheels, in *IEEE Power Engineering Society General Meeting*, June 6–10, 2004, Denver, CO. IEEE Press, Piscataway, NJ, pp. 2038–2042.

313. Gomis-Bellmunt, O., Liang, J., Ekanayake, J., and Jenkins, N. (2011) Voltage–current characteristics of multiterminal HVDC–VSC for offshore wind farms. *Electric Power Systems Research*, **81**, 440–450.

314. Díaz-González, Bianchi F.D., Sumper, A., and Gomis-Bellmunt, O. (2014) Control of a flywheel energy storage system for power smoothing in wind power plants. *IEEE Transactions on Energy Conversion*, **29**, 204–214.

315. Hu, W., Chen, Z., Wang, Y., and Wang, Z. (2009) Flicker mitigation by active power control of variable-speed wind turbines with full-scale back-to-back power converters. *IEEE Transactions on Energy Conversion*, **24**, 640–649.

316. Sørensen, P., Hansen, A.D., and Carvalho-Rosas, P.A. (2002) Wind models for simulation of power fluctuations from wind farms. *Journal of Wind Engineering and Industrial Aerodynamics*, **90**, 1381–1402.

317. Tascikaraoglu, A., Uzunoglu, M., Vural, B., and Erdinc, O. (2008) Power quality assessment of wind turbines and comparison with conventional legal regulations: a case study in Turkey. *Applied Energy*, **88**, 1864–1872.

318. GAMS, http://www.gams.com/ (accessed May 8, 2012).

319. Skogestad, S. and Postlethwaite I. (2007) *Multivariable Feedback Control, Analysis and Design*, John Wiley & Sons, Ltd, Chichester.

320. Ruiz-Álvarez, A., Colet-Subirachs, A., Álvarez-Cuevas, F., *et al.* (2012) Operation of a utility connected microgrid using an IEC 61850-based multi-level management system. *IEEE Transactions on Smart Grid*, **3**, 858–865.

321. Lazarewicz, M.L. (2011) *Status of Flywheel Storage Operation of First Frequency Regulation Plants*. Beacon Power Corporation, Wilmington, MA.

322. Brown, P.D., Peças Lopes, J.A., and Matos, M.A. (2008) Optimization of pumped storage capacity in an isolated power system with large renewable penetration. *IEEE Transactions on Power Systems*, **23**, 523–531.

323. Bueno, C. and Carta, J.A. (2006) Wind powered pumped hydro storage systems, a means of increasing the penetration of renewable energy in the Canary Islands. *Renewable and Sustainable Energy Reviews*, **10**, 312–340.

324. Chen, G.Z, Liu, D.Y., Wang, F., and Ou, C.Q. (2009) Determination of installed capacity of pumped storage station in WSP hybrid power supply system, in *International Conference on Sustainable Power Generation and Supply: SUPERGEN '09*, April 6–7, 2009, Nanjing, China. IEEE Press, Piscataway, NJ, pp. 1–5.

325. Goel, P.K., Singh, B., Murthy, S.S., and Kishore, N. (2009) Isolated wind–hydro hybrid system using cage generators and battery storage. *IEEE Transactions on Industrial Electronics*, **58**, 1141–1153.

326. Grant-Wilson, I.A., McGregor, P.G., and Hall, P.J. (2010) Energy storage in the UK electrical network: estimation of the scale and review of technology options. *Energy Policy*, **38**, 4099–4106.

327. Agbossou, K., Kolhe, M., Hamelin, J., and Bose, T.K. (2004) Performance of a stand-alone renewable energy system based on energy storage as hydrogen. *IEEE Transactions on Energy Conversion*, **19**, 633–640.

328. González, A., McKeogh, E., and Gallachóir, B.Ó. (2003) The role of hydrogen in high wind energy penetration electricity systems: the Irish case. *Renewable Energy*, **29**, 471–489.

329. Ipsakis, D., Voutetakis, S., Seferlis, P., *et al.* (2009) Power management strategies for a stand-alone power system using renewable energy sources and hydrogen storage. *International Journal of Hydrogen Energy*, **34**, 7081–7095.

330. Little, M., Thomson, M., and Infield, D. (2007) Electrical integration of renewable energy into stand-alone power supplies incorporating hydrogen storage. *International Journal of Hydrogen Energy*, **32**, 1582–1588.

331. Nakken, T., Strand, L.R., Frantzen, E., *et al.* (2006) The Utsira wind hydrogen system – operational experience. *Proceedings of the EWEC. 2006*, http://www.ewec2006proceedings.info/ (accessed May 7, 2013).

332. Onar, O.C., Uzunoglu, M., and Alam, M.S. (2006) Dynamic modeling, design and simulation of a wind/fuel cell/ultra-capacitor-based hybrid power generation system. *Journal of Power Sources*, **161**, 707–722.

333. Cavallo, A. (2007) Controllable and affordable utility-scale electricity from intermittent wind resources and compressed air energy storage (CAES). *Energy*, **32**, 120–127.

334. Lund, H., Salgi, G., Elmegaard, B., and Andersen, A.N. (2009) Optimal operation strategies of compressed air energy storage (CAES) on electricity spot markets with fluctuating prices. *Applied Thermal Engineering*, **29**, 799–806.

335. Swider, D.J. (2007) Compressed air energy storage in an electricity system with significant wind power generation. *IEEE Transactions on Energy Conversion*, **22**, 95–102.

336. Barote, L. and Marinescu, C. (2009) A new control method for VRB SOC estimation in stand-alone wind energy systems, in *International Conference on Clean Electrical Power*, June 9–11, 2009, Capri, Italy. IEEE Press, Piscataway, NJ, pp. 253–257.

337. Dufo-López, R., Bernal-Agustín, J.L., and Domínguez-Navarro, J.A. (2009) Generation management using batteries in wind farms: economical and technical analysis for Spain. *Energy Policy*, **37**, 126–139.

338. Brekken, T.K.A., Yokochi, A., von-Jouanne, A., *et al.* (2011) Optimal energy storage sizing and control for wind power applications. *IEEE Transactions on Sustainable Energy*, **2**, 69–77.

339. Lex, P. and Jonshagen, B. (1999) The zinc bromine battery system for utility and remote area applications. *Power Engineering Journal*, **13**, 142–148.

340. Anagnostopoulos, J.S. and Papantonis, D.E. (2008) Simulation and size optimization of a pumped-storage power plant for the recovery of wind-farms rejected energy. *Renewable Energy*, **33**, 1685–1694.

341. Caralis, G. and Zervos, A. (2010) Value of wind energy on the reliability of autonomous power systems. *IET Renewable Power Generation*, **4**, 186–197.

342. Kapsali, M. and Kaldellis, J.K. (2010) Combining hydro and variable wind power generation by means of pumped-storage under economically viable terms. *Applied Energy*, **87**, 3475–3485.

343. Papaefthimiou, S., Karamanou, E., Papathanassiou, S., and Papadopoulos, M. (2009) Operating policies for wind-pumped storage hybrid power stations in island grids. *IET Renewable Power Generation*, **3**, 293–307.

344. Bernal-Agustín, J.L. and Dufo-López, R. (2008) Hourly energy management for grid-connected wind–hydrogen systems. *International Journal of Hydrogen Energy*, **33**, 6401–6413.

345. Sherif, S.A., Barbir, F., and Veziroglu, T.N. (2005) Wind energy and the hydrogen economy – review of the technology. *Solar Energy*, **78**, 647–660.

346. Zafirakis, D. and Kaldellis, J.K. (2009) Economic evaluation of the dual mode CAES solution for increased wind energy contribution in autonomous island networks. *Energy Policy*, **37**, 1958–1969.

347. Joerissen, L., Garche, J., Fabjan, Ch., and Tomazic, G. (2004) Possible use of vanadium redox-flow batteries for energy storage in small grids and stand-alone photovoltaic systems. *Journal of Power Sources*, **127**, 98–104.

348. Hu, W., Chen, Z., and Bak-Jensen, B. (2010) Optimal operation strategy of battery energy storage system to real-time electricity price in Denmark, in *IEEE Power and Energy Society General Meeting*, July 25–29, 2010, Minneapolis, MN. IEEE Press, Piscataway, NJ, pp. 1–7.

349. Castronuovo, E.D. and Lopes, J.A.P. (2004) On the optimization of the daily operation of a wind-hydro power plant. *IEEE Transactions on Power Systems*, **19**, 1599–1606.

350. Loisel, R., Mercier, A., Gatzen, C., *et al.* (2010) Valuation framework for large scale electricity storage in a case with wind curtailment. *Energy Policy*, **38**, 7323–7337.

351. Carton, J.G. and Olabi, A.G. (2010) Wind/hydrogen hybrid systems: opportunity for Ireland's wind resource to provide consistent sustainable energy supply. *Energy*, **35**, 4536–4544.

352. Korpas, M. and Greiner, C.J. (2008) Opportunities for hydrogen production in connection with wind power in weak grids. *Renewable Energy*, **33**, 1199–1208.

353. Denholm, P. and Sioshansi, R. (2009) The value of compressed air energy storage with wind in transmission-constrained electric power systems. *Energy Policy*, **37**, 3149–3158.

354. Hennessy, T. (2006) *Overcoming Transmission Constraints: Energy Storage and Wyoming Wind Power*, http://www.sandia.gov/ess/docs/pr_conferences/2006/farber.pdf (accessed May 7, 2013).

355. Nyamdash, B., Denny, E., and Malley, M.O. (2010) The viability of balancing wind generation with large scale energy storage. *Energy Policy*, **38**, 7200–7208.

356. Hida, Y., Yokoyama, R., Shimizukawa, J., *et al.* (2010) Load following operation of NAS battery by setting statistic margins to avoid risks, in *IEEE Power and Energy Society General Meeting*, July 25–29, 2010, IEEE, Minneapolis, MN. IEEE Press, Piscataway, NJ, pp. 1–5.

357. Iijima, Y., Sakanaka, Y., Kawakami, N., *et al.* (2010) Development and field experiences of NAS battery inverter for power stabilization of a 51 MW wind farm, in *International Power Electronics Conference (IPEC)*, June 21–24, 2010, Sapporo, Japan. IEEE Press, Piscataway, NJ, pp. 1837–1841.

358. Lee, T.-Y. (2007) Operating schedule of battery energy storage system in a time-of-use rate industrial user with wind turbine generators: a multipass iteration particle swarm optimization approach. *IEEE Transactions on Energy Conversion*, **22**, 774–782.

359. Chang, Y., Mao, X., Zhao, Y., *et al.* (2009) Lead–acid battery use in the development of renewable energy systems in China. *Journal of Power Sources*, **191**, 176–183.

360. Roberts, B.P. (2008) Sodium–sulfur (NaS) batteries for utility energy storage applications, in *IEEE Power and Energy Society General Meeting – Conversion and Delivery of Electrical Energy in the 21st Century*, July 20–24, 2008, Pittsburgh, PA. IEEE Press, Piscataway, NJ, pp. 1–2.

361. Butler, P.C., Cole, J.F., and Taylor, P.A. (1999) Test profiles for stationary energy-storage applications. *Journal of Power Sources*, **78**, 176–181.

362. IEEE Power & Energy Society (2007) *IEEE Standard 1013–2007: IEEE Recommended Practice for Sizing Lead–Acid Batteries for Stand-Alone Photovoltaic (PV) Systems*. IEEE, New York.

363. IEEE Power & Energy Society (2007) *IEEE standard 1562–2007: IEEE Guide for Array and Battery Sizing in Stand-Alone Photovoltaic (PV) Systems*. IEEE, New York.

364. Kyocera Solar (2015) *Datasheet for KC200GT PV Module*, http://www.kyocerasolar.com/assets/001/5195.pdf (accessed May 25, 2015).

365. Gradella-Villalva, M., Gazoli J.-R., and Ruppert-Filho, E. (2009) Comprehensive approach to modeling and simulation of photovoltaic arrays. *IEEE Transactions on Power Electronics*, **24** (5), 1198–1208.

366. Joint Research Centre – European Commission (2015) *Interactive Map on Photovoltaic Geographical Information System*, http://re.jrc.ec.europa.eu/pvgis/apps4/pvest.php (accessed May 25, 2015).

Index

Energy Storage in Power Systems, First Edition. Francisco Díaz-González, Andreas Sumper and Oriol Gomis-Bellmunt.
© 2016 John Wiley & Sons, Ltd. Published 2016 by John Wiley & Sons, Ltd.